Prof. Cobezon:

Gratitude!

Roy

5. 6. 2024

Co-Existing
with the Earth

Tzu Chi's Three Decades of Recycling

Co-Existing
with the Earth

Tzu Chi's Three Decades of Recycling

William Kazer
Rey-Sheng Her

World Scientific

NEW JERSEY · LONDON · SINGAPORE · BEIJING · SHANGHAI · HONG KONG · TAIPEI · CHENNAI · TOKYO

Published by

WS Lifestyle, an imprint of
World Scientific Publishing Co. Pte. Ltd.
5 Toh Tuck Link, Singapore 596224
USA office: 27 Warren Street, Suite 401-402, Hackensack, NJ 07601
UK office: 57 Shelton Street, Covent Garden, London WC2H 9HE

Library of Congress Control Number: 2020949541

British Library Cataloguing-in-Publication Data
A catalogue record for this book is available from the British Library.

Cover image designed by the Tzu Chi Foundation.

CO-EXISTING WITH THE EARTH
Tzu Chi's Three Decades of Recycling

ISBN 978-981-123-223-7 (hardcover)
ISBN 978-981-123-157-5 (paperback)
ISBN 978-981-123-158-2 (ebook for institutions)
ISBN 978-981-123-159-9 (ebook for individuals)

For any available supplementary material, please visit
https://www.worldscientific.com/worldscibooks/10.1142/12141#t=suppl

Typeset by Stallion Press
Email: enquiries@stallionpress.com

Foreword I

"Learning Crucial Lessons from a Crisis: Sincerely Repent and Show Gratitude to Mother Nature"

By Dharma Master Cheng Yen

Tzu Chi has been promoting environmental protection for three decades. When it began its recycling campaign 30 years ago, few people saw environmental protection as an urgent task. Since then, much has changed. Now, environmental protection is a huge global issue. Our planet Earth appears to be running a fever. Due to human actions of senseless destruction, our climate is out of balance and natural disasters occur at an alarming rate. Global bodies like the United Nations have responded to this unfortunate turn of events. The UN works tirelessly to resolve these grave problems by promoting energy conservation and reducing industrial pollution and carbon emissions. But the task of protecting the environment is an arduous one and the ultimate goal remains elusive, largely due to the endless desires of humankind. Preserving the environment depends on achieving a common understanding and working together towards a shared purpose.

It is apparent to all there has been an increase in the occurrence of natural disasters due to the enormous effects of greenhouse gases. It is

Ven. Dharma Master Cheng Yen, Founder of Buddhist Tzu Chi Foundation.

equally clear that if we continue to consume energy at the current pace, there will be even more calamities in the years ahead. Even with the benefit of this knowledge, we are unable to reach a common understanding and agree on the necessary response. No matter how urgently we work to avert this crisis, people still prioritize economic gain. This is indeed very worrying. How long can planet Earth withstand this ceaseless consumption and the resulting pollution? We need to raise our consciousness. More importantly, we must have a consensus on taking collective action. Climate change is here. Temperatures are climbing, the ice caps covering the North and South Poles are melting and sea levels are rising. These are cause for great concern and sadness.

How do we protect planet Earth? Tzu Chi has some of the answers. All along we have been promoting conservation and the preservation of resources. Since I first called for protecting the environment by "turning the hands of applause into the hands of recycling" some

30 years ago, tens of thousands of volunteers — many of them senior citizens — have joined the cause. This large group of volunteers now has a vast knowledge of recycling. With just a touch of the hand, our "Recycling Bodhisattvas" know what can or cannot be recycled and they have become skilled at sorting these materials properly.

In some cases, our elderly volunteers may lack formal education, but they know the value of true sincerity. Every day, they put these virtues to work, adhering to traditional values and ethics. They have worked hard their whole lives, raising a family and contributing to society, whether it was building roads and bridges, cultivating a rice paddy, or running a street-side food stall. It is especially remarkable that our volunteers continue to work so diligently in the cause of recycling even after their hard-earned retirement. They use their calloused hands to care for planet Earth by engaging in recycling. These hands are indeed the most beautiful ones.

But there are limits to what we can do through recycling alone. We need to change our habits, reflect upon our actions more profoundly, reduce our daily desires and do the right thing. We need to live more frugally by returning to a simpler lifestyle of the past and reducing our carbon footprint.

I offer you a small example from my own experience. In the past, when I wrote for the *Tzu Chi Monthly*, instead of wasting a clean sheet of paper, I would use the back of a personal calendar as a scratch pad. I would first use a pencil to write on it, then a blue pen the second time, next a red pen and finally a writing brush. A sheet from a simple calendar could be used four times. That is why I feel sad when I see young people using large amounts of paper without the slightest thought for the consequences of their wasteful behavior.

How long is a lifetime and how many things can one enjoy in that lifetime? However long, there are only 24 hours in a day. However much pleasure we seek, we can only eat so much and only wear so many clothes. If we cherish the blessings we have, we accumulate blessings. When others throw things away, they are discarding their "blessings." If we pick up these "blessings" we can accumulate them through recycling. We can turn these recycled materials into something useful to help those who are suffering.

When there is a tug of war between good and evil, the side with more people has the greater strength. If there are more kind-hearted people, more people willing to conserve and protect the environment, then there will be enough strength to save the Earth. If there are more wasteful people engaging in evil deeds, this will only accelerate the destruction of our planet. Therefore, we must work harder and appeal to others to see the importance of environmental protection and actively contribute to this cause.

We should learn to be content with our surroundings. The Earth accommodates our needs and provides us with a bountiful existence. But it seems like it is never enough, as we constantly inflict damage on our environment in our quest for more. We have endless desires and limitless greed. This is why crises multiply and intensify. Humankind has not yet awakened from its slumber. We see horrific disasters caused by an imbalance among the four elements of earth, air, fire and water.

Now we face a different kind of disaster as the coronavirus continues to spread among us, causing much pain and suffering. We grieve for the victims as the staggering death toll from this pandemic continues to mount. We are fearful amid a frantic search for a cure. And we don't really know the origin of this affliction. Nonetheless, our fears are not helpful. The pandemic is indeed a great calamity, but we can learn from this tragic experience. We can transform distress into gratitude because this is a "lesson from a great calamity."

Humankind must wake up and help restore our natural harmony. We must sincerely repent and show gratitude towards Mother Nature. Humans rely on this verdant planet. Trees and grass are the vitality provided by nature; grains and water are the nourishment given to all.

To restore the natural balance, we must "respect all life with gratitude and love." Humans are not the only living beings on planet Earth; animals are also living creatures too. Raising animals for humans to consume is selfish, wasteful and unsustainable. It requires vast tracts of land to provide large amounts of feed. The land needed to satisfy our demand for meat is directly related to the amount of waste and pollution produced. If demand for meat can be restrained, there will be a reduction in the need to raise livestock. That in turn will curb the emission of greenhouse gases, which will ease the strains from global warming and slow the pace of climate change. The reduced need for slaughtering animals will allow all living things to multiply, grow and die according to natural law. That will avoid the bad karma acquired from satisfying one's own taste buds. The less bad karma created from enmity and hatred, the greater the blessings of love and goodness. Only then can calamities be mitigated.

There are far too many disasters in the world right now. We need to ensure a safe and peaceful future. We need to stop the rise in temperatures and stabilize the climate. The way to achieve these goals is through vegetarianism and recycling.

Ultimately, we must cherish the Earth and all sentient beings. This is what love is. When we are able to love the things we have, then we are able to love others. I hope that everyone can inspire love and cherish Mother Nature. Let love infuse our thoughts and actions so that we can co-exist with the Earth and all its inhabitants.

Foreword II

Disaster Risk Reduction and Sustainable Development

The following foreword is based on an edited transcript of remarks made by Dr. Rajendra K. Pachauri, the renowned Indian environmentalist and former chairman of the United Nations Intergovernmental Panel on Climate Change (IPCC), via a video link at the Fifth Tzu Chi Forum in Taiwan in September 2019. Under his leadership, the IPCC was co-awarded the Nobel Peace Prize in 2007 (with Albert Arnold (Al) Gore, Jr.). Dr. Pachauri, who passed away in February of 2020, was also the Chief Mentor of the Protect Our Planet (POP) Movement and founder of the Lighting a Billion Lives initiative, which aims to bring clean energy to the rural poor.

I'd like to talk to you all about climate change, which is by far the most serious challenge that human society is facing. I was vice chairman of the IPCC (Intergovernmental Panel on Climate Change) for five years and then chairman for 13 years from 2002 to 2015, during the course of which, I had the privilege of receiving the Nobel Peace Prize on behalf of the Panel. And that clearly was a turning point because it provided the scientific community with the inspiration, with the encouragement that has propelled us forward to ensure that human society receives the gift of knowledge which assesses the climate change that's taking place in all these aspects. And during the period that I was chair of the IPCC and in subsequent years we have certainly been able to create a huge of amount of awareness of all over the world.

Unfortunately, there are some leaders and (people) in positions of power who still believe, as a result of vested interests perhaps, that climate change is not for real. But we know now that it is for real because in 2011 the IPCC brought out a special report on "Extreme Events and Disasters," and to my mind, that was a game changer because prior to that everybody believed climate change is just a warming of the whole system, and it doesn't matter if you change the temperature of the earth by 2 or 3 degrees. But I think that particular report clearly brought out the fact that climate change is accompanied by an increasing frequency and intensity of extreme events.

What are these extreme events? Heat waves, extreme precipitation events, extreme sea level-related events, and of course, because of heat waves, because of increasing temperature, we also have increasing forest fires. You would be aware that this year itself, in the Arctic region, there were fires ranging from Siberia to Alaska. Now that is unheard of. At the same time, we have rapid melting of the body of ice in Greenland. I have been to Greenland myself. It's a huge, solid amount of ice about 3 kilometers high, and you can imagine what could happen if some of that ice was to melt and go into the oceans. We will have sea level rises which will be unprecedented and would change the entire geography of this planet.

Now this is why I would like to mention that Taiwan is an island, and there are many islands across the globe which are going to face huge hazards in terms of sea level rise. And that, to my mind, is clearly one of the most serious impacts of the climate change, quite apart from others that would affect agriculture, that would affect the availability of water and even human health.

So, what we need to do is not only to mitigate the emissions of greenhouse gases but also make sure that we adapt to the growing impacts of climate change. Now in order to mitigate the emissions of greenhouse gases, what we really need to do is to move beyond the use of fossil fuels. The IPCC brought out a special report on 1.5°C which clearly specifies that the world has to stop the increase of warming to no more than 1.5°C by the end of this century because between 1.5°C and 2°C there are impacts that would be completely unbearable and that would certainly make an enormous difference in terms of risks for human society and all living species.

More recently the IPCC also brought out a special report on "Climate Change and Land," and this clearly specifies the fact that human society cannot consume meat in the manner it has done all these years. What we really need to do is to shift to plant-based diets. Now I did say this in 2007 at a press conference in Bangkok. I addressed the media and told them that human society has to move to consumption of much lower quantities of meat, and I said you would be healthier and so would the planet. Of course, people ridiculed me. And a year later, I addressed a gathering in London, and that's when the mayor of London (Boris Johnson), who is now the Prime Minister of the UK, wrote a paper ridiculing me in a newspaper.

And this (view), of course, was prevalent at that time because people don't want to give up the ways that they have been following all these years.

Fortunately, there was a city called Ghent in Belgium where I addressed another gathering. And the mayor and the citizens of that city introduced one day of a week as a meatless day.

And then Sir Paul McCartney and I went to the European Parliament to appeal to them that one day a week in Europe should be a meatless day, but of course nothing happened in response to that.

So, what I want to do is to request the Tzu Chi Forum and this remarkable organization to join hands with me and make sure that we move human society towards plant-based diets.

And if we can do that, I am sure this would have huge benefits to human society and for the eco-systems of this planet.

I also want to mention that I have two major organizations, called the "World Sustainable Development Forum" and the "POP Movement" which stands for "Protect Our Planet." And with the latter, what we are trying to do is to mobilize youth all over the world so that they start taking action; they become the leaders of grassroots initiatives by which human society can move towards a carbon neutral future and to limit temperature increase to 1.5°C. We would have to make sure that by 2040, human society becomes carbon neutral.

So, I want to appeal to you all and to make sure that we join hands because I think dealing with climate change requires collective action. I think you are a motivated and enlightened group, and if we can join hands together, then I am sure we can go a long way. I think we can create a sustainable future.

So, once again, I offer my homage. I offer my salutation to the Master. And I would like to wish you all the best. I am sure this Forum would be extremely productive and very useful. I am sorry to be missing it, but I am grateful to you for giving me this opportunity to address you in this panel.

Thank you very much!

Foreword III

Recycling and Environmental Protection 30th Anniversary — Walking on the Path of Sustainability

By Mr. Po-Wen Yen, CEO of Taiwan Buddhist Tzu Chi Foundation

Tzu Chi has been advocating recycling for 30 years but Tzu Chi's life-style of protecting the environment has an even longer history. It began when Dharma Master Cheng Yen first led her monastic and lay disciples in adopting a self-sufficient Zen farming lifestyle at the Jing Si Abode. This lifestyle embodies the spirit of simplicity and frugality, the concept of appreciating and cherishing all things, as well as living in harmony with the Earth. These fundamental principles completely correspond to the 5Rs lifestyle promoted by current environmental protectionists, namely: recycle, repair, reuse, reduce, and refuse.

That was why Master Cheng Yen could not bear to see the huge increase in the amount of trash produced in Taiwan after years of rapid economic growth and greater material wealth.

On 23rd August 1990, during a talk on "Living a Blessed Life" at Shin Min Commercial and Industrial Vocational High School in Taichung City, Master Cheng Yen was interrupted by an enthusiastic round of applause from the audience. Master Cheng Yen took the opportunity to address public cleanliness and the environment by describing her reaction to something she saw on a morning walk: piles of rubbish left behind at a shuttered market. She made an appeal to the audience to use those same hands that had just applauded her remarks and turn them to the task of protecting the environment. This was a critical first step for many Tzu Chi volunteers on their journey towards collectively protecting the environment.

For the past 30 years, Tzu Chi's environmental protection volunteers have never ceased their efforts to preserve our Earth, making their daily rounds in countless neighborhoods and carrying out recycling collection work, 365 days a year, regardless of conditions.

From Taiwan and extending across the globe, there were 110,000 environmental volunteers working in 10,000 recycling stations and environmental protection education stations in 19 countries as of the end of 2019. These volunteers work tirelessly every day, using their hands to protect the planet.

Tzu Chi's Recycling and Environmental Protection Mission is not only focused on recycling. Instead, following Dharma Master Cheng Yen's teachings of "Purity begins at the source," Tzu Chi environmental protection volunteers do recycling and they also learn to purify their hearts by reining in wants and desires and reducing consumption and waste.

Tzu Chi has established a comprehensive model of environmental protection. It embodies a complete circular economy from the sorting of recyclables collected by volunteers, to the integration of eco technology in the fabrication of disaster relief supplies and lifestyle products.

Funds generated by recycling are channeled to sustain the operations of Da Ai TV, which instills kindness and humanity in its programs, touching the hearts of many who in turn become environmental protection volunteers. This circular relationship creates a second cycle: one of love from Tzu Chi's humanitarian spirit. By joining these two circles, the symbol of infinity ∞ is formed. It becomes a continuous process which is sustainable on its own and which encourages a never-ending flow of people to support its cause through environmental actions.

Every year, KPMG interviews 1,300 CEOs of major corporations, including 30 from Taiwan, regarding their perspective on the future outlook. In the 2019 KMPG Global CEO Outlook, all CEOs, both Taiwan and globally, listed "Environmental Issues and Climate Change" as the main risk factor in future developments.

Extreme weather events and natural disasters are irrefutably brought on by climate change. People across this planet Earth face the risk of becoming displaced or ending up as climate refugees in the near future. As humans continue to overconsume our natural resources and destroy the balance of the environment, the Earth is fighting back. Only by taking action to peacefully "co-exist with our Earth" can we lessen the effects of this climate emergency.

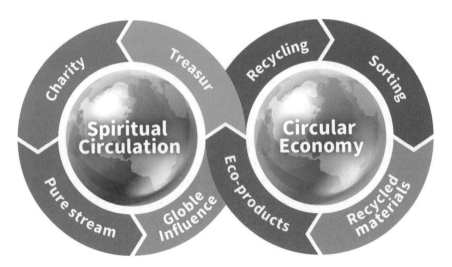

Environmental Protection-Tzu Chi Model (Chart provided by Tzu Chi Foundation)

In recent years, Tzu Chi has been actively involved in the NGO platforms of the United Nations, sharing and promoting its experience in "turning rubbish into gold, and the gold into love" with the aim of inspiring individuals to take action, starting with their own daily lives. In 2019, Tzu Chi held 49 international dialogues and events and was invited to speak collaboratively at 37 events. Recognized by the international community for our efforts, Tzu Chi is honored to be an NGO observer in the United Nations Environment Programme (UNEP). It also participated in the sixth session of the Committee on Disaster Risk Reduction, the Caritas International 21st General Assembly, the High-Level Political Forum on Sustainable Development (HLPF), and the United Nations Framework Convention on Climate Change (UNFCCC)'s conference of the parties (COP). This has given Tzu Chi the opportunity to promote recycling, a zero waste circular economy, plant-based diets, the use of non-disposable cutlery and much more.

The year 2020 marks the 30th anniversary of Tzu Chi's Environmental Protection mission and the 50th anniversary of Earth Day. In such a year of great commemorative value, Tzu Chi launched its 30th anniversary event themed "Eco 30 Global Actions" with a series of activities including eco forums, environmental education exhibitions, an "eco run" in collaboration with National Geographic, and zero waste picnics to promote sustainable living.

In November, "The Sixth Tzu Chi Forum" on "Future Earth and Green Initiatives" was held at the Jing Si Hall in Xindian-Taipei. Representatives from the industrial, government, and educational sectors, as well as environmental protection groups, were all invited to participate and examine ways to deal with the current emergencies of global warming and climate change, by employing the Master's appeal to the whole world for "shared understanding, consensus, and collective action." A collective response is essential to avert crisis and ensure sustainable living for humanity.

To mitigate the risks of climate change, Tzu Chi actively promotes two broad areas of sustainability:

1) Sustainable resources: promoting a circular economy to increase the recycling of resources and working toward the goal of zero waste.
2) Sustainable green planet: Promoting the spirit of "purity at the source, returning to simple living," a plant-based diet and the 5Rs. These actions can prevent the spread of disease, improve individual health, protect the environment, and mitigate the effects of climate change with the aim of living sustainably with our Earth.

With the help of this book, I hope to present the outstanding environmental protection achievements of the combined efforts of the Taiwan government and non-governmental organizations, as well as provide a record of Tzu Chi's dedication to environmental protection, and its process of implementation, over the past 30 years. Through this book, we hope to exchange ideas across various sectors and use the power of our shared environmental direction to achieve a sustainable future.

Preface

The Tzu Chi Foundation has gained renown as a Buddhist charity with a deep social commitment. It has won the admiration and gratefulness of many over the years and has been a significant force for good in disaster relief. In many cases it has been the first responder, quickly handing out food and financial assistance, providing shelter and medical aid while offering psychological support to those in distress. From its start in Taiwan in 1966, it has extended its hand of compassion across far flung parts of the world; now more than 100 countries have benefited from its relief efforts. While I was aware of its important role in this sphere, I was rather late in discovering Tzu Chi's significant contribution to environmental protection and recycling in Taiwan and elsewhere.

My enlightenment came several years ago, while I was researching an article for a Taiwan government publication on the environment and what is known as the circular economy. It was then that I finally gained an appreciation of Tzu Chi's role in recycling and environmental protection. I interviewed a senior official at the Environmental Protection Administration for the official view of Taiwan's effort and its assessment of what more was needed. It was a productive and wide-ranging discussion, but it clearly was only one part of a much bigger story. The government could guide public behavior through its policy directives, but it needed public cooperation on a broad scale. I asked

William Kazer and Tsai Kuan, 102-year-old Tzu Chi environmental volunteer. (Photographer / Lo Shi-Ming)

this official for recommendations for other people and organizations to talk to, and without skipping a beat, he replied that I needed to talk to people at Tzu Chi. Any report on this subject without their input would be incomplete, he said. He then went on to praise Tzu Chi's organizational skills and suggested that I would be amazed by what I saw.

I can say that I took his advice and his high praise for Tzu Chi was not misplaced. What impressed me was not only the extent of Tzu Chi's organizational capability but also the selfless dedication of so many of its volunteers. I wondered how Tzu Chi could motivate so many people to work together for the benefit of the community. What was its secret formula? In the years since, I have gained a modest understanding of this impressive organization and I can say without reservation that my admiration has only grown.

This book is an effort to give readers a glimpse of the invaluable contribution that Tzu Chi has made to environmental protection in Taiwan and beyond and how its efforts have touched so many lives in the process. I also hope the readers will see what is behind this campaign which has spread to so many other countries. And in some small way I hope it helps these tens of thousands of volunteers continue their invaluable work. Special thanks to Shi-Ming Lo for his invaluable editorial assistance.

William Kazer

Contents

Introduction

The Karma of Zero Waste

Thirty years ago, the Buddhist Compassion Relief Tzu Chi Foundation began its recycling mission out of a conviction that economic progress was taking too heavy a toll on the environment. Taiwan was considered to be one of the "Asian tiger economies" at that time, recording robust economic growth and boosting living standards at a rapid pace. While economic progress brought greater well-being to the island's residents, it left environmental degradation in its wake and encouraged wasteful lifestyle habits that needed to be changed. Thirty years ago, Taiwan's environment truly was in need of help. Taiwan had the unflattering reputation of being a "garbage island" as a result of its breakneck economic development at the expense of the environment. A throwaway lifestyle was particularly inappropriate on a relatively small island like Taiwan which was literally running out of space to bury all the waste.

Tzu Chi's founder Dharma Master Cheng Yen concluded it was no longer possible to sit by idly and watch the degradation of the environment and the depletion of its resources. It was time for action. Mother Earth needed help.

Tzu Chi's recycling awakening came in response to some soul-searching questions: What would our planet look like if we continued our wasteful ways? Don't we all have a shared responsibility in

Master Cheng Yen calls on her followers to join a campaign to promote recycling. Speaking at a gathering in 1990 she urged her listeners to turn their hands to the work of environmental protection. (Photo provided by Tzu Chi Foundation)

protecting the Earth for future generations? What can we do to change things?

It was Master Cheng Yen who inspired this mission that ultimately helped launch Taiwan on a path to becoming a global example for recycling and environmental protection. In 1990, she had an epiphany, if you will, while on her way to speak to her followers in the central Taiwan city of Taichung. As she passed by the shuttered stalls of a local marketplace, she surveyed the unsightly detritus left behind after the close of business: discarded paper and empty bottles carpeting the street and used plastic bags clogging drains or fluttering in the wind like airborne declarations of waste and neglect.

Master Cheng Yen's initial reaction was one of disgust, but in this brief, unpleasant moment she found a new and inspiring message to bring to her audience. Acknowledging applause for her prepared remarks, she asked if perhaps those same hands that had just clapped so enthusiastically could be turned to recycling. The task ahead was a daunting one — to save the Earth.

For the tens of thousands of members of the Buddhist group this was a call to action. Tzu Chi is a charity organization dedicated to serving the community in the broadest sense and promoting humanistic values. So, pitching in to help Taiwan and the world at large was something that came naturally to its members, who embrace the organization's goals and values.

Tzu Chi volunteers started small. They combed their neighborhood looking for recyclable materials. They used their own homes or found space nearby to set up temporary recycling stations. They separated paper, plastic bottles, and glass containers from trash. But they discovered there was a lack of understanding of what could and what should be recycled. There was too much trash mixed in with the recyclable materials. Well-meaning neighbors were eager to pitch in, but they needed more guidance. They brought in all sorts of cast-offs, like old mattresses and worn-out shoes with kitchen waste and even dead animals thrown into the mix. Volunteers needed to make it clear that they were not collecting trash and that the objective was to retrieve materials that could be repurposed. They had to demonstrate what was recyclable and what was not. Slowly the message got through. Tzu-Chi volunteers gradually expanded the recycling campaign, and the selfless dedication of the volunteers eventually convinced others to join in. Today there are more than 89,000 recycling volunteers and 8,000 recycling stations in Taiwan alone. This collective effort has also been expanded to more than 19 countries and territories.

How is it that this movement has been able to spur so many people into action? What is the secret sauce? If you ask members of Tzu Chi, it is a combination of factors.

First is that the members of Tzu Chi have had years to become acquainted with the concepts espoused by Master Cheng Yen. Her message of simplicity and social responsibility have inspired trust,

After sorting and cleaning, used aluminum cans gain new value as recycling material. (Photographer / Shu, Ze-Ren)

according to Jason Leou, a Tzu Chi senior staff specialist and commissioner. (A Tzu Chi commissioner has completed a training course and an internship and agrees to follow a code of behavior as well as meet other requirements). "The key is Master Cheng Yen," he says. "You need one person to lead. So many people trust her judgment and so many people have adopted her values. They share her philosophy and have confidence in the success of this recycling effort."

That message is echoed time and time again. Kan Wan-Cheng, who first joined as a volunteer 28 years ago, says that like many others, he too was inspired by Master Cheng Yen. He first heard her speak when he was a 15-year-old and gradually took a greater interest in Tzu Chi and recycling. "In the early days we were engaged in simple recycling operations in local communities. We recycled plastic bottles, metal cans and paper. I went about it with a sincere attitude and with the passing of time I gained a deeper understanding. I saw this become a seed planted around the world."

Timing may also have played a role. By the mid-1990s the cost of Taiwan's profligate ways was becoming increasingly obvious. "The garbage issue was becoming very apparent," says Mr. Leou. Add to that the moving stories of selfless behavior of early volunteers who took action on their own for the greater good, often donating recycling proceeds to promote the cause. These factors all contributed to the resounding success. "It was a virtuous circle," he says.

As Tzu Chi's recycling campaign gained momentum, it also delivered a critical social dividend by ensuring that other resources — human ones — were used to the fullest. As tens of thousands of volunteers were mobilized in this enormous task, many of them found a new purpose in life and a welcome bond with their fellow recyclers.

Unfortunately, in our materialistic society, many people, young and old alike, have become convinced they have nothing to contribute. They become depressed or withdrawn, and sometimes they lose hope. But recycling and preserving the environment for the next generation

Volunteers Chen Chiu-tsu and Hsu Wen-ping prepare used plastic bags for recycling. (Photographer / Huang Xiao-Zhe)

is a noble cause that can stimulate a sense of purpose. It can demonstrate that even those who had once despaired indeed have value. It builds a sense of community as participants work toward a common purpose. And in many cases, Tzu Chi's volunteers overcome serious physical handicaps to make their own contribution.

Many of the volunteers at the recycling stations are senior citizens who have long since reached the retirement age; some are undeterred by health problems or physical limitations. They find fulfillment in doing work that helps protect the environment, and they often become healthier, both mentally and physically, in the process (see Chapter 4).

Master Cheng Yen likes to say that serving as a volunteer in the recycling drive is a way to overcome arrogance and learn humility. Squatting beside piles of refuse and sifting through mounds of empty, unwashed plastic bottles to find useful raw materials can indeed be a humbling experience. But it is an effort that forges strong bonds among people who share common goals. Volunteers say participating in this community effort is a way to demonstrate their sense of gratitude for their own good fortune. In some cases, this may reflect their recovery from serious illness or injury or perhaps a personal setback in their daily life. They may feel a need to give something back to their neighbors. In this sense we see recycling as a much broader mission — while showing an appreciation for the Earth, the volunteers have found a new sense of self-worth. This could be seen as a recycling of the individual human spirit.

One of Tzu Chi's most remarkable assets is its ability to communicate the need to protect the environment and inspire so many individual citizens to personally get involved in recycling work. Volunteers absorb this guiding principle and embrace it in their daily lives. Many of these same people become grassroots teachers, sharing their environmental knowledge with others at home and in their community. Those who have participated in recycling understand the importance of conserving the Earth's resources and promoting an environmentally friendly lifestyle.

Who says work can't be fun? Recycling volunteers enjoy their mission of "saving the Earth." (Photographer / Huang Shi-Ze)

Do they believe that it is part of making merit and producing good karma that has positive consequences in the future? Perhaps. But they clearly do understand that the reduction of waste is an obligation to future generations and a form of selfless compassion.

Tzu Chi's recycling infrastructure can also serve as a platform for society as a whole in this effort. This is reflected in the organization's close cooperation with government departments, corporate executives, schoolteachers, students, and the general public (see Chapters 8 and 9).

Over the years, Tzu Chi has also devoted much of its energy to providing relief to disaster victims and helping them repair their damaged lives. In many instances Tzu Chi has been the first non-governmental relief group to reach a disaster area. Moreover, it has often continued to aid disaster victims years after the call for help was first answered. This has allowed Tzu Chi to establish enduring bonds with these stricken communities. Over the years, the trust it has earned has allowed it to impart an awareness of the need to protect the environment and minimize our carbon footprint. Global warming and climate change have added to the frequency and intensity of natural disasters and protecting the Earth has become a key theme of Tzu Chi's disaster prevention efforts. Tzu Chi works to strengthen public awareness of the need to recycle and protect the environment as part of a global effort to prevent and mitigate disaster.

Tzu Chi's recycling efforts can be said to provide the following benefits:

• reducing waste, improving recycling quality, boosting recycling income;
• empowering community members to join in sorting recyclable materials;
• promoting environmental awareness among the public;
• inspiring more people to join the effort to promote environmental protection;
• promoting ethical volunteering and charitable acts;
• cherishing the elderly and engaging seniors in meaningful activities;
• strengthening community responsibilities and neighborhood relationships;
• offering positive free-time activities for families and groups.

In the following pages we will examine how these benefits have been attained.

Chapter 1

Tzu Chi's Philosophy and the Environment

Personally Experience the Unity of the Environment and All Beings

Any movement or ideology that spreads its influence far and wide cannot attribute its success solely to the uniqueness of its contemporary nature. A more important aspect is whether the ideology is robust and has goals that are sincere and can be embraced on an individual level. Tzu Chi's environmental mission in Taiwan has all of these attributes. It has gradually expanded into the global environmental protection movement, which covers a full range of objectives including saving energy, reducing the emission of greenhouse gases, adopting environmental technology, and promoting recycling. Its orientation is as an environmental ideology promoted through the practice of Master Cheng Yen.

Tucked away in the quiet township of Sincheng in Taiwan's Hualien County is the Jing Si Abode, where Master Cheng Yen started her spiritual cultivation. It is also the place where the Buddhist Tzu Chi Compassion Relief Foundation was established. Its home is a simple yet elegant structure with a white roof set against the dark green backdrop of the majestic Central Mountain Range. It looks out over the endless open expanse of the Pacific Ocean.

1

The Jing Si Abode in Hualien, Taiwan (Photographer / Bai, Kun-Ting)

Aesthetics aside, no one knows exactly why Master Cheng Yen chose to settle in this place. Some people say that when she reached Sincheng township, she saw a Bodhisattva statue that was left by the Japanese. The image of the statue looked like the face of the Bodhisattva that was in a dream she had in her youth. At the time, her mother had been struggling with a lengthy illness. The young Cheng Yen had a dream that a Bodhisattva handed her medicine which she then gave to her mother. She experienced the same dream over three consecutive nights. And in less than two weeks, her mother recovered.

The combination of the mountains and the sea, a humble abode and a group of Buddhist nuns, formed a kind of solemn harmony. Through the ages, Chinese philosophers have spoken of the harmony between heaven and man. Chinese artists have painted dramatic landscapes showing majestic mountains and flowing rivers, depicting humans almost as an afterthought. Humans are barely perceptible, only to be found in hidden corners, suggesting the insignificance of mankind in relation to the grand sweep of nature. China's philosophers never describe the incorporation of humans in a heaven specially designed for them and situated in a particular place. Instead, there is a unity of heaven and man as a concept more akin to poetic and artistic feelings — almost like a realm beyond the senses of those of us in the real world.

Master Cheng Yen decided to settle in this county where heaven, mountains and a great ocean come together in harmony to start her personal quest for life's enlightenment. It is here that she found a bright, spiritual ray of light. And in less than 50 years, this ray of light crossed the high peaks of the mountains to shine across all of Taiwan and then beyond the immense ocean towards distant shores. Master Cheng Yen has a very Chinese view of environmental interaction, stressing the need to "respect the environment and attain harmony between heaven and man." Heaven, Earth, and all people blend into each other in a sense of communication and harmony. "With a quiet heart, listen to the Earth breathe," she says.

Master Cheng Yen prayed daily, chanting the "Lotus Sutra" alone in her small wooden hut, maintaining an ascetic lifestyle. She'd go to bed at midnight, wake up at two in the morning and make an offering to the Buddha. There are three volumes of the "Lotus Sutra" which the Dharma Master studied. One was the "Sutra of Innumerable Meanings," and it had a particularly deep connection with her lifelong adherence to Buddhism. In the silence of the night, when the young practitioner was praying, one segment of the sutra was particularly memorable. "Attain ultimate calmness so the mind is crystal clear and reflects all things. The wish is great and without limit yet still humble. Keep this state of mind permanently with all innumerable dharmas learned. With great wisdom, we come to know all universal law." This was an enlightenment and inspiration, according to Master Cheng Yen. Years later she described the experience to her disciples as a "reflection of her inner self emerging, as if her heart and the entire universe converged, bringing a revelatory comprehension."

In that tiny hut beneath the mountains and surrounded by a serene ambience, the darkness of the night covered the endless land and evening stars lit the distant sky. The beginning and the end of the universe seemed to converge in this space that measured less than ten square meters. The young practitioner quietly contemplated the significance of the scriptures. Her profound understanding was that when the mind is totally at rest, desires are silenced. The calmness of the heart was like the still surface of a peaceful lake that reflects everything clearly. So, infinite wisdom emerged naturally, beneath the pure spirit, attaining comprehension of great wisdom that has access to all dharma and is embodied in all scriptures. This is an indication of a practitioner's achieving oneness with the universe. This is not poetic or mystical, however. It was achieved through the thorough experience and insight into an inner self. Its practice affects those who follow her footsteps in the future.

Under the majestic mountains and the vast landscape, Master Cheng Yen was able to connect the infinite wisdom of the universe with the nature of human wisdom. This comprehension corresponds with the enlightenment of the Buddha under the fabled Bodhi tree.

We can imagine the contrast between the majestic mountain, the vast ocean and the small hut. The practitioner blends into our consciousness and its comprehension into our wisdom. This comprehension and reflection form a beginning. This brings out an awareness of mankind and our surroundings. It also reveals a new path for our future behavior. Master Cheng Yen's concept of the environmental ambience has Chinese characteristics, with the paradox of harmony between heaven and man and the interdependence of man and the Earth. This is also a Buddhist view of nature, emphasizing the love of life, the equality of all things and the notion that all beings have no differences. Master Cheng Yen brought "love" into the environmental view by emphasizing the love of Mother Nature, the need to cherish all things and cherish all matter. She also stressed the spirit of modern scientific technology with a pragmatic point of view that brings about a specific idea of practising environmental co-existence. This is possible in everyone's daily life and it signifies that all things are equal.

Man as Part of Nature

In fact, the relationship between man and nature is the most important foundation of human existence. If we look at the Old Testament, man and nature were originally united as one. In the Garden of Eden, man and nature, man and woman, man and all beings were all in harmony. There was no choice, no freedom, no thoughts until Adam and Eve ate the forbidden fruit from the tree of knowledge, which led to the sense of shame and the beginning of man's detachment from nature as a whole. Once man leaves the whole natural environment, he is destined never to be united with all of nature again in its mysterious yet perfect state.

Social psychologist Erich Fromm in his book *Escape from Freedom* points out that once man departed from nature, although he was able to attain individual freedom, he became aware of his loneliness and fear. Because man cannot tolerate this loneliness and fear, in his subconsciousness he hopes to return to nature and resurrect this unity with all beings. It is like an infant leaving its mother's womb to start its independent life. What follows is the feeling of loneliness and pain. Fromm said: "To be back with nature; to be back in the embrace of their mother became an eternal need of man."

Why do people need to be loved? From Fromm's point of view, it is because to love somebody or to be loved gives man a feeling of regaining the unity of the soul. Love became a formal way of regaining unity after man was forcibly detached from nature.

The biblical story of Adam and Eve and the theft of the fruit of knowledge in the Garden of Eden seems to explain how Western society seems to be stuck with the tensions and confrontational relationship between man and nature. In contemplative Christianity, God and man are not equal; God is the master of mankind. The rage of

God will punish mankind for its evils (such as the partaking of the forbidden fruit and disobeying God) while those who follow his righteous path will be rewarded. The existence of a supreme being and his relationship with man is a connection between the dominant and the dominated. Contemplative Christians gradually edged towards the scientific theory that emerged with the Renaissance period; the relationship between the supreme being and man — or nature and man — corresponded to the structure of what later became known as the capitalist society. This eventually turned into a scramble of exploitation and confrontation. The relationship of the supreme being and his domination of man changed when man realized that he could dominate the supreme being through the force of science and thereby control his own destiny. Man comes from nature, but he believes that he can use science and civilization to dominate it and all beings as well.

The Chinese treatment of nature does not exhibit the same level of tension or confrontation seen in the West. Chinese hope to co-exist with nature, and this can be seen in the work of Tao Yuan-ming, a poet of the Jin Dynasty period (265–420 AD). His poem "To Drink" offers these dreamy lines: "Picking chrysanthemums by the eastern fence, gazing leisurely towards the southern mountains; the mountain air is fresh at dusk, birds fly back and forth; in these things there is great wisdom, but no words are needed to express it." This relationship with nature is something that is innately understood. In the *Tao Te Ching* the great Chinese sage Lao Tzu wrote: "Man follows the land; land follows heaven; heaven follows the Tao and the Tao follows nature." It directly points to the existence of man within an eternal cycle and universal law.

Confucius had this to say on a similar theme: "Heaven commands that man follows nature's four seasons; everything under the sun should follow heaven's command." Heaven is nature, everything comes from nature; man should follow the law of nature and its seasons in order to survive.

The influential religious scholar Mircea Eliade (1907–1986) pointed out that the ancient Chinese originally used the word "emperor" to refer to dominating everything in the universe using natural forces. (Eliade M., 1989/translated by Dong Qiang, 2001). From the start of the Shang and Chou dynasties, Chinese emperors insisted on emperor worship, which was a way to demonstrate respect to the gods in heaven, or as a gesture in the hope of attaining unity with the gods. That is why the emperors were called the "son of heaven." Emperors were considered to be the masters of heaven on earth. Dynasties promulgated decrees that stated they ruled "in the name of heaven." Heaven was the foundation for dynastic survival, indicating the inseparable relationship between man and nature, or heaven and earth.[1]

Displays of care for the environment rarely appear in documents or books of traditional Chinese intellectuals. This may be because it is our modern day industrial society that has caused most of the environmental damage. In the distant past, man's contribution to natural disasters was more commonly viewed in the context of superstitious explanations for calamitous events.

Tzu Chi is a deeply contemplative extension of Buddhism and other aspects of Chinese civilization. Master Cheng Yen uses her life encounters and creativity to re-interpret the relationship of man and nature with actual practice by both individuals and as a group. When she says "walk like you are afraid to hurt the land," that gives life to the land itself. She also says: "Turn your hands to the work of environmental protection." This encourages thousands of volunteers to actively participate in protecting the environment. It gives value to waste, safeguards the environment and has a lasting impact on each individual's view of life. This is a group and individual effort as a force for transcendence.

---

[1] M. Eliade, Histoire des croyances et des idées religieuses: de l'âge de la pierre aux mystères d'Eleusis, trans. Dong, Qiang (Paris: Payot, 1989; Taipei: Shang-zhou Publishing Co., 2001), 262.

A Quiet Symphony

Each morning at 4 o'clock, before the first rays of light appear, the bells and drums of Jing Si Abode announce the start of the morning class. From a distance, the solemn chanting of the nuns can be heard, together with the chorus of birds and frogs and the rustling of leaves in the wind. It is like a quiet symphony that echoes softly in the time and space between the dreams of night and the waking hour.

Standing in the morning stillness, you can feel and almost hear the beat of your own pulse. It is as if the trees that line the sides of the Jing Si Abode are listening to the sacred chants. For an hour or more, some 300 nuns and volunteers are immersed in peaceful meditation together. Then, as the bird calls grow louder, the world seems to rise from its slumber to greet the new day.

Perhaps it was on a morning like this that Master Cheng Yen first felt the never-ending breathing of the Earth. "Listen to the breathing of the Earth with a silent heart," she says. The breathing of the Earth can actually be heard. Once you experience meditation at the Abode, you will be able to feel such pure, quiet solemnity.

Although the atmosphere of the Jing Si Abode at this early hour is calm and reverent, the daily routine is hardly relaxed. After breakfast the nuns set about their daily chores. They have a conviction that "no work for a day is no food for a day." They grow fruits and vegetables and tend the garden. They grind barley powder for use as a traditional food additive and take turns with the communal cooking. They make candles and other handicrafts for sale to visitors. They carefully place their kitchen waste in a compost heap beside the rows of papaya trees for use as fertilizer. They shun actions that harm the Earth, taking Zen and Buddhism as their guide to the chores of farming. These ordinary and seemingly repetitive routines can bring about a sweetness of life. They prove that life does not require outside stimulation to be extraordinary. An awakening can be achieved by searching inside one's own self.

China's Zen masters maintained that cutting firewood, fetching water and performing other chores are all part of Zen. Real Zen is attained not only through meditation, when one contemplates in silence and reaches a state of calmness. Real silence and calmness can be attained through a variety of small actions no matter when or where the practitioner is. It could be in addressing a small situational change, for example. Whatever action is being taken, one remains calm. This exercise is referred to as "calm in motion."

In the early years, Master Cheng Yen would till the fields using the traditional method — with the help of a water buffalo. When the animal refused to cooperate, she would extend a handful of grass to coax it along until the task was completed. (Whipping — or even a gentle switching — to prod an animal along was to be avoided). Weeding is also part of the regimen. February is the coldest month in Hualien and on one particularly chilly February day the Dharma Master was set to weed the Abode's big field but found it hard to move her cold fingers. She started with a small section of the field where she could get her reluctant hands around the stubborn weeds. Soon she was able to work her way through an expanding portion of the field. Eventually she accomplished her task. This is part of the Tzu Chi ethos — whether it is in charity or relief work or other duties: Start small and move forward from there. The Dharma Master was able to learn simple but important lessons from her days working the field. No matter whether it was tilling the soil or saving sentient beings, the Master Cheng Yen applied the same principles. And in doing this she used love as her guiding light.

The nuns may have rough hands from their work, but their inner self is tranquil and trouble-free. Talking to them reveals the joy of the dharma and the wisdom brought about by love and tolerance. In the early days, the nuns had to follow Master Cheng Yen in working the fields and caring for the fruit trees. They were fewer in number so the load was heavier. They also needed to read the Buddhist scriptures as well as the ancient Confucian classics. This is the formula for creating a complete personality, through farming and a bond with nature. The frugal surroundings help build the character of these practitioners.

This simple characteristic of self-denial is the source of integrity that is essential in the real world.

Nature is the mother of mankind. So why is it that man worries about the possible destruction of Mother Earth? The main threat is from the greed of human beings. To produce more and feed our growing appetite we continuously expand our farming areas, burn forests, and fell trees. But in our rush to expand croplands we unleash destructive forces that threaten our environment. Author Jared Diamond, who wrote the book *Collapse: How Societies Choose to Fail or Succeed* noted that Easter Island was at one time considered a paradise. Its civilization was able to sculpt and transport huge stone statues that are still considered to be wonders of the world. But population pressure led to the destruction of forests, the extinction of many plant species and the disappearance of vital resources. Excavation of kitchen waste has revealed human bones that confirm the island had faced famine. This great civilization ultimately lapsed into cannibalism.

Diamond also warned us not to take the extinction of the Easter Island civilization as an isolated case. Humanity in the vast universe is like an isolated island, and as an island, its resources eventually may run out. His conclusion: Do not let the world become the next Easter Island. A self-disciplined lifestyle is the answer to avoiding devastation — such as from the greenhouse effect — which could result in the collapse of the world as we know it today.

Cherish Everything, Start with Self-Discipline

Seven thousand years ago, the ancient shamans buried the remains of the bears they hunted in the belief that the bears would be reborn.[2] Ancient people respected their prey. If they hunted on one side of the forest, they would not hunt in the same area the next season. The

---

[2] M. Eliade, *Histoire des croyances et des idées religieuses: de l'âge de la pierre aux mystères d'Eleusis*, trans. Dong, Qiang (Paris: Payot, 1989; Taipei: Shang-zhou Publishing Co., 2001), 45–47.

ancients knew that they had to give nature time to restore itself. But with the onset of industrialization, the balance changed. A highly industrialized society began disruption on a wide scale. Man was able to slaughter livestock in great numbers by mechanical means. This is cruelty on a massive scale. In some pig farms, pigs are confined in tight cages from birth. They are unable to stand until the day they are slaughtered. Gill fishnets catch all sizes of fish from the sea, threatening marine resources. Since mankind's industrial development began some three centuries ago, more than 50 million whales have been killed. Billions of animals are slaughtered each year just to feed humans. The demand for more cropland means the rain forests face endless encroachment. Man's unlimited greed inflicts ever greater damage on nature. This eventually could lead to our self-destruction through the greenhouse effect.

Master Cheng Yen has said that in order to save planet Earth, we must live a frugal life guided by our inner self. The nuns in the Abode set a good example. Damage to the environment will not stop unless we adopt a frugal lifestyle. We must experience the value of material things. We must convert our respect for nature into action.

The Jing Si Abode is a very frugal temple. The Tang-style main building measures less than 60 square meters. It is used as a place for chanting the Buddhist texts, holding conferences for charity work, and communicating with Tzu Chi members. In the early years, it was a dormitory and dining hall as well. Everything was done here. Even after more than 10 renovations, the Abode maintains a maximum height of three floors. This is the spiritual home for every Tzu Chi member from all over the world.

Self-denial and thrift are the core values at the Abode, which is the character embodied in its working environment. The wooden lodge with its window latticework has a spartan look to it. Simple wooden furniture is used in the offices and visitors' room. Most rooms use ceiling fans instead of air-conditioning. It is difficult for people who come here from faraway places to imagine that this is the center of one of the world's largest charitable organizations.

Tzu Chi members tend a garden at the Jing Si Abode. (Photographer / Chen Zhong Hua)

At the Jing Si Abode, there is an endless stream of volunteers and visitors from all over the world. The nuns need to accommodate these travellers, and the kitchen is busy preparing meals for more than 400 people each day. The funds come from the hardworking nuns who support the Tzu Chi Foundation. It is concrete proof of the fact that the Tzu Chi religious arm is not a dependent of the charitable foundation. Instead, it is a donor.

In one corner of the Jing Si Abode there are several nuns mending clothing and quilts for other nuns. As much as possible, they take care of their daily needs themselves. Whatever they can save, they save. Besides growing much of their own food, they fry peanuts to extract oil for cooking. They produce a kind of hard and salty beancurd. One small piece can be served with two bowls of rice. This kind of dish is a reverential nod to earlier days when life here was even more challenging.

Meals are served on round tables, giving the dining room a warm family atmosphere. There are serving spoons and separate chopsticks for serving purposes. Whatever is left will be served to the next batch of people who come to dine. Food is placed on one's small plate and everyone is supposed to finish whatever is taken. There is a small pot of water at the table. Everyone drinks from a bowl and washes the small plates clean. This kind of eating ritual has been practiced for more than 40 years not only by the nuns, but also by more than a hundred colleagues working in the Abode.

There is a candle room at the back of the Abode, which is packed with nuns and volunteers who come to help every day. The nuns melt wax into candle molds made from empty plastic bottles that once contained Yakult brand yoghurt drinks. After the wax cools slightly, the wick is inserted. This is called the "candle without tears," as the design keeps the wax from dripping. It saves wax and gives the plastic container a new purpose. It is a design that is cherished by the Dharma Master for its small role in protecting the environment as well as for the pleasant light given off when the candle flickers in the dark. This inspires contemplation and it symbolizes the giving of light to make others happy but without the shedding of tears. Happiness is helping others without harming oneself.

Master Cheng Yen's life has always been simple and rigorous. She firmly believes that everything has a hidden life that is rich. Long ago, when writing letters to her friends, she often used the back of letters from those same friends for her reply. She insists that a piece of paper has a life; it has to be used to the full. The Dharma Master uses only one bucket of water a day, emphasizing the importance of conserving water.

She dines from a simple wooden table that has been with her for more than 20 years. She sits in a small rattan chair that is assumed to be more than 20 years old too. Everything is simple, unadorned and useful. Frugality, simplicity and industriousness are the common traits seen in the daily lives of the Dharma Master and the nuns. They use natural light in the visitor's room. Electric lights are in use only when there are documents that need to be read. The Dharma Master's study room is very small with only a table lamp at her desk. After 40 years, life is much like it was in the little wooden hut, with the exception that electric lights are now available when needed.

A spartan and frugal life is not only a virtue; it is also the source of energy for Mother Earth. A society that overexploits its resources will ultimately suffer from self-destruction. A frugal lifestyle is a manifestation of respect and affection for nature. Man must control his desires in order to attain the real coexistence of man and nature. In Tzu Chi's way of thinking, this is possible only through real practice.

The spirit of the Abode is the same spirit that is seen in Tzu Chi volunteers all over the world in their poverty relief efforts. It is the spirit they display in bringing relief when disaster strikes. Tzu Chi members are taught to be thrifty so that they are not placing even more pressure on the environment. They practice self-denial to

One of our neighbors at the Jing Si Abode. We are not the only living beings on planet Earth. We must respect all life. (Photographer / Xu Rong-Hui)

protect mankind and save planet Earth. The so-called "butterfly effect" suggests that a butterfly flapping its wings in Beijing might cause a large hurricane on the other side of the world. Its central theme is that the behavior of each human being affects every other being. Applying this same theory, wastage on one side of the world could be the reason for hunger and famine on the other side. In the history of mankind, the excessive consumption of resources is the culprit behind climate disturbances such as the floods and droughts we are experiencing today. Those who are unable or unwilling to control their desires are contributing to global warming. They will be compounding the problem of rising sea levels and the sinking of coastal cities in the years to come.

Material Things Have the Same Value as My Life

Master Cheng Yen has said: "Every wriggling spirit has its Buddha's nature." This has been Master Cheng Yen's feeling towards living beings and all natural resources, be it water, soil or air.[3] The beings that Buddhist doctrines mention do not refer to human beings alone; they include anything that is either with or without life.

To love the land is different from enjoying the beauty of the forest or experiencing a quiet early morning breeze. The Dharma Master wants Tzu Chi's members to use their labor to personally experience the love of the land and use their hands in demonstrating their love of the earth. This emotional attachment to matter is expressed in the surroundings of every Tzu Chi building. All the open spaces are covered with connecting porous cement blocks which are used in place of cement or asphalt in the hope that the land can breathe. This is a reflection of Master Cheng Yen's devotion to preserving our land.

On September 21, 1999 a devastating earthquake left a trail of destruction in Taiwan. Tzu Chi was quick to respond in handing out

---

[3] Shih De-fan, *Cheng Yen Shangren Nalu Zuji* (Taipei: Tzu Chi Cultural Publishing Co, 2006), 70.

We must live in harmony with Mother Nature. (Photographer / Bai, Kun-Ting)

immediate relief and it was a key participant in the reconstruction effort. Tzu Chi built 51 schools in the aftermath of that tragedy. But the Dharma Master's attention was focused not only on restoring disrupted lives and ameliorating suffering. She tried to establish a new humanitarian treatment of the land. More than 200,000 volunteers helped in the landscaping of the schools built by Tzu Chi. Students, parents and Tzu Chi volunteers from all over Taiwan helped create green spaces in the hope that the children would be able to grow and flourish in a natural environment. All buildings were designed with open corridors that run along galleries that reveal the sky and green mountains. A subtle gray color is used for the buildings to highlight the blue sky and green plant life. The design is simple but elegant, emphasizing frugality.[4]

All the school grounds use cement bricks instead of asphalt or concrete so that the Earth can "breathe" and rainwater can seep back into the soil below. This not only saves water it allows the trees to be nurtured as well. All 50 schools of Tzu Chi's Project Hope use these cement bricks. Every Tzu Chi construction, including hospitals, schools and its own headquarters, use a total of more than 5 million cement bricks in the open spaces.

Each hospital that Tzu Chi has built over the years has these cement bricks put in place before the hospital's opening day. We often see groups of Tzu Chi volunteers in blue and white uniforms laying bricks in front of the medical center. These include directors, assistant directors, department heads and doctors squatting on the ground, using their hands to help put the bricks in place. They brave the hot sun to accomplish their task. These doctors are in fact treating a different patient on these days; they are healing the sick Earth.[5]

And there is another way for Tzu Chi to help heal the sick Earth. The entire Tzu Chi organization adheres to a vegetarian diet. Being a vegetarian reduces the stress on our environment from the raising of livestock to feed our human population. Moreover, it is a way to practice dharma. Tzu Chi's fundamental philosophy is to respect and cherish all lives. Animals convey life as much as human beings. Adhering to a vegetarian diet reminds us of the basic Buddhist tenet that all sentient beings are equal and we should treat them equally with love and compassion.

---

[4] Liu Feng-juan, Xie Lei-nuo, Tu Miao-yi, Fan Yu-wen and Ruan Yi-zhong, "Special Report on the 3rd Anniversary of 921 Earthquake," *Tzu Chi Monthly* 430 (2002): 10–57.

[5] Xie Ming-jin and Jian Ling-jie, "Let's Build a Hospital Together," *Ren Yi Xin Chuan* 36 (2006): 82–84.

Master Cheng Yen once said that even though animals cannot express their feelings to us, nonetheless, when they face slaughter, their fear, anger, and hatred emerge. We then swallow all these negative feelings. She believes that it is enormously harmful to our body and spirit. Not to kill is good for our body, our soul and the planet. It is said that you are what you eat. If so, by eating violence, we might cling to violence.

Chapter 2

Tzu Chi's Recycling and the Volunteer Spirit

Master Cheng Yen has been advocating the practice of Buddhism since her early years. She counsels Tzu Chi volunteers that all things are equal; we are part of a sentient world that includes each and every creation. That is why we must love all sentient beings and cherish life. In her view, the purpose of life is to practice Buddhism, while every location is a temple that affords the opportunity to experience it. Recycling is a way of spiritual and physical cultivation; it embodies the concrete practice of loving all sentient beings. Moreover, through recycling, Tzu Chi volunteers not only extend the lifespan of all things, they also preserve the value of life itself. This is truly using the recycling station as a spiritual temple.

Tzu Chi's environmental mission originated on August 23, 1990 when Master Cheng Yen gave a public lecture where she called on her followers to join hands and support an ambitious recycling mission. While the humble activity of collecting garbage has helped protect the environment, it has also transformed the inner soul of many of those who joined the effort and helped them reaffirm the value of life. What is the relationship between these recycled materials and the people who extend their usefulness? Can the love shown in recycling help people rebuild an individual's self-confidence and self-esteem?

Chinese philosophers have said that forgetting both matter and self can unite mind and matter. But mind and matter are difficult to reconcile philosophically, not to mention scientifically. Let us look to the French sociologist Pierre Bourdieu for a possible explanation. "When we are exposed to certain things, whether physically or socially, we will attain a certain temperament (disposition). The individual agent develops these dispositions in response to the objective conditions they encounter."[1] From his viewpoint, when the mind (or heart, if you prefer) and matter come into contact, some kind of sense and blending can be attained.

If Bourdieu's insights are correct, when the heart and matter interact, there is a mutual effect. This may explain why, through Tzu Chi's recycling efforts in treasuring material life, we experience the transformation of our own selves for the purification of our soul. Whenever a volunteer picks up a PET bottle or a piece of paper, does his inner self reach a higher realm? Look at the Tzu Chi volunteers, who diligently pick up a discarded plastic bottle or a piece of paper casually thrown on the ground. This common experience is uplifting and proves that man and material can form a bond, that "material life" and "life" itself are inseparable.

The Value of "Useless" Material

"Useless objects are very useful." This is the most profound wisdom that volunteers receive from recycling. This is not only value as a measure of material wealth, but also as it affects the soul.

As in Bourdieu's estimation — what he called the "habitus theory" — the human body can even change a person's temperament directly. As the volunteers watch "useless" junk being recycled every day, their temperament eventually undergoes a change. They discover that if a discarded plastic bottle or used piece of paper can be recycled to protect the earth and purify the mind, how about his old but still active physical body?

---

[1] Pierre Bourdieu, *Outline of a Theory of Practice* (Cambridge and New York: Cambridge University Press, 1977), 72, 95.

In the course of recycling, the volunteers are enlightened by their observation of the new use for what seemed like useless objects. As their physical body changes within this recycling environment, this changes their attitude towards life too. Recycling strengthens their sense of value and helps them gain new respect in their community. In many ways, they experience the feeling of rebirth through recycling.

One such person was Chang Lin Chao of Chiaohsi township in Yilan. She started recycling at the age of 90 and carried on until she passed away at the age of 104.[2] People from her village wondered why she took on this challenge so late in life. Despite the doubts expressed by people around her, she persisted. Gradually, her determination influenced her neighbours to join her. She became the village's best-known recycling volunteer, and thanks to her efforts, the sanitation of the community greatly improved. Through recycling, her neighbors learned that pollution harms Mother Earth.

Recycling discarded materials has far-reaching effects. In the process of recycling, older volunteers saw the creation of new value from items once considered useless. A piece of wastepaper could be used to make pulp while a plastic bottle could become a blanket and copper filament could be turned into wire. How about the volunteers? These old, "useless" bodies seemed to have some value. The psychological cues and the strengthening of self-worth give many volunteers a sense of rebirth. Through recycling, they are convinced that they still have value. They build their own value by enhancing cleanliness in their own communities as guardians of the earth.

The Equality of Life

Physical activity can change a person's temperament. This has proved true for many Tzu Chi recycling volunteers who are disabled. Recycling

---

[2] Shan-hui shu-yuan, *Cheng Yen Shangren de Nalu Zuji*, vol. Spring (Taipei: Tzu Chi Cultural Publishing Co, 2007), 274–287.

Chuo Su-Hwei has serious vision problems but that hasn't stopped her from joining in the recycling effort. Here she is being interviewed by Da Ai TV. (Photo provided by Tzu Chi Foundation)

makes them feel as if they are whole again, as a person with value in their life.

The Buddhist scriptures offer the parable of a blind man threading a needle. Aniruddha, one of the Ten Disciples of the Buddha, dozed off when the Buddha was lecturing. His inattention was noticed by the Buddha and this caused Aniruddha great discomfort and regret. He was so disturbed by this that he refused to close his eyes, even while sleeping. As a result, he eventually went blind. But despite his blindness, he was extremely diligent. While he was blind physically, he was able to sew his own clothes. On one occasion his thread fell from the needle and he asked for someone to help him re-thread it. The Buddha was meditating nearby, and, with compassion, helped Aniruddha re-thread his needle.[3] This story makes the point that physical disabilities can be overcome while it also illustrates the compassion of the Buddha. In the Tzu Chi recycling stations, we see similar examples.

Chuo Su-Hwei of the Bagua recycling station in Kaohsiung seems to be a modern day version of this story. She has had poor eyesight since her childhood. She can only see light and shadow in one eye, while she is completely blind in the other. Still, she is able to play a part in the recycling effort, dismantling electric fans and stacking old newspapers.[4]

The Truth We Practice

Russian psychologist Lev Vygotsky wrote in *The Zone of Proximal Development*: "The truth in our lives comes from the truth that we practice." This is much like the experience of the recycling volunteers. The true value of life is expressed in the recycled cans, paper and PET bottles. The recycling experience was internalized and confirmed by the volunteers. "Like all functions of consciousness, it originally arises from action," Vygotsky said. Real construction of a concept comes not from language, but from practice.[5]

This is why Master Cheng Yen describes practice as a temple. It is a framework for internalizing truths and values. Countless

---

[3] Master Cheng Yen, *The Innumerable Sutra*, Sutra Collection-7 (Taipei: Tzu Chi Cultural Publishing Co, 2001), 121–123.

[4] Cheng Yen Shangren, "Life, Transform a Moment to Eternity," *Tzu Chi Monthly*, 432 (2002): 6.

[5] L. S. Vygotsky, *Mind in Sociology* (Cambridge, Massachusetts: Harvard University Press, 1978), 9–30.

volunteers — some of them grappling with physical or emotional challenges — have experienced life changes thanks to recycling. Despite these challenges, they are able to project extremely positive energy and gain the admiration of those around them.

No scripture or statement can serve as a substitute for practice. Master Cheng Yen said that, in Buddhist scriptures "all moving objects have a Buddha's nature"[6] while sentient beings include those with or without life or shape. The Chinese concept of "unity of mind and matter" is just an abstract philosophy, but, when environmental volunteers use their hands to touch recycling materials, they realize the precious value of every life. This is a real experience, and this experience reflects the philosophy and values of life that "the mind and matter are one" and "all matter has emotions."

Esteem is Transferred through Love

The same kind of experience may produce different results. For example, a commercial trash collector who is mainly motivated by profit might have a different concept of the value of recycling than a Tzu Chi volunteer. At Tzu Chi, love is passed on and strengthened during the recycling process.

Tzu Chi recycling stations provide a feeling of "family love" — something that has long been lost in the modern world. The recycling stations are like a traditional society where villagers dry their harvest in the village courtyard in the daytime and sit around together and chat under the stars at night. In the old village system, there was mutual help. There was a sense of love and value that went hand in hand. Without love, the concept of value is cold. It can even become dogmatic. When the essence of love is not found in value, then, everything is presumed as "should," and "should" will be mostly about the system

---

[6] Shan-hui shu-yuan, Cheng Yen Shangren de Nalu Zuji, Vol. Spring (Taipei: Tzu Chi Cultural Publishing Co, 2007), 274–287.

that governs our relationships. The psychoanalyst Karen Horney called this a "should atrocity." But when things are done through the medium of love, it makes people happy and inspires them to join in. Value is like sheet music; it can make happy music dance in the depth of our souls.[7]

Ask any environmental volunteer why they like to work in the recycling station. Almost all answer they feel a sense of joy. This inexplicable sense of joy is the value they earn, as well as the great love that brings happiness to this group.

This sense of a big family is felt not only by those who join in as volunteers. Those who reject recycling initially may become inspired by this family atmosphere later on.

An environmental volunteer described the process of how he got started in recycling work. He said he was invited by neighbors to help drive the environmental vehicles. He thought: "Well, I'll just help out." When his delivery was completed, he was ready to go. But just as he wanted to leave, a volunteer ran in and said, "Everybody! Come and have some snacks!" He thought: "Okay. Eat something and leave." Once he had a little refreshment, he wanted to leave but felt embarrassed and stayed to work for a while longer. Just as he was about to leave, another volunteer said: "Come and drink some winter melon tea!" Again, he could not get away. He ended up spending the whole day at the recycling station. But surprisingly, he said he felt a tremendous sense of accomplishment and inexplicable joy that day.

A doctoral candidate in biological sciences at Academia Sinica, a renowned Taiwan research institution, relates a similar tale of how he was attracted to Tzu Chi's recycling operations. He started going to the Neihu recycling station in Taipei to wait for his wife who was participating in Tzu Chi activities. He thought to himself that, instead of waiting around on the sidelines, he might as well help out with the

---

[7] Karen Horney, "The Tyranny of the Should" in *Neurosis and Human Growth*, trans. Li, Ming-bin (New York: Norton, 1950; Taipei: Zhiwen Publishing Co., 1976), 64–65.

recycling. He had not been particularly aware of Tzu Chi's environmental protection program, but he found that there were many professionals and successful entrepreneurs at the recycling station. Everyone there treated him like a son. He was able to learn from the life experiences of the older volunteers he met. He likened this to the great love of a big family. He eventually set aside time to join in recycling on a regular basis, as part of his spiritual education.

A recycling station is like a home, filled with love and positive energy. The warm feeling of home is what attracts so many people. Once they start recycling, their concept of environmental protection slowly builds.

Recycling and Physical Rehabilitation

Many modern ailments are a result of the lack of love, caused by loneliness and a sense of worthlessness. Psychiatrists have long confirmed that a person's mental state will determine his physical state. Many environmental volunteers, after working in the recycling station, miraculously find their weak or sick body regaining health and vitality. Sociologist Chris Shilling, author of *The Body and Social Theory*, has pointed out that the human body is a constantly changing physical and social phenomenon.[8] Why is the human body social in nature? Sociologist Jon Ivar Elstad noted that those at the bottom rung of society easily become sick. Social life, self-esteem, a sense of security, self-control, as well as a feeling of being loved, maintain a person's health status.[9]

It is easier to become ill in an environment of low self-esteem or lacking social support. Conversely, a person's body is in a better state if he or she has loving support, is welcomed by others, or is engaged in work that has a profound significance. Sociologist Peter Freund stated:

---

[8] Chris Shilling, *The Body and Social Theory* (London: Sage, 1993), 37–61.
[9] J. I. Elstad, "The Psycho-social Perspective on Social Inequalities in Health," *Sociology of Health and Illness*, 20, no. 5 (1998): 598–618.

"The interaction of the structural context of social life or social contact has a direct impact on the physical state of the person."[10]

These theories explain why many Tzu Chi recycling volunteers experience remarkable physical changes when they perform environmental work. Recycling stations provide every environmental volunteer a network of love, as well as the force of warm and solid social support. This force is what medical sociologists point to as the cause of their greatly improved physical and mental state.

Hsu Chin-lian, who is in her eighties, had a badly bent spine when she joined the volunteers at the Neihu recycling station. After doing recycling work for a year, her back gradually straightened to a discernible extent.[11] At work at the recycling station, it was as if she had forgotten her physical pain. She would move boxes around and fill them with recyclables. Eventually, she was no longer complaining of her old aches and pains. "When my back straightened, I went back to my family home and everyone was amazed! Carrying out environmental protection work really makes people healthier, and happier," she said.

It was not just the exercise from recycling that enabled the physical improvement. The environment of warm family values was also a factor.

Rescuing Distressed Souls

"Situational education" is one of the key components to a lifetime committed to purifying the mind, according to Master Cheng Yen. A person's transition through situational or group involvement is what Master Cheng Yen has described as "situational education."[12] With

---

[10] P. E. S. Freund, "The Express Body: A Common Ground of Sociology of Emotion and Health and Illness," *Sociology of Heath and Illness,* 12, no. 4 (1990): 452–477.

[11] Shan-hui shu-yuan, *Cheng Yen Shangren de Nalu Zuji* (Taipei: Tzu Chi Cultural Publishing Co., 2006), 42.

[12] Shan-hui shu-yuan, *Cheng Yen Shangren de Nalu Zuji*, Vol. Spring (Taipei: Tzu Chi Cultural Publishing Co, 2007), 274–287.

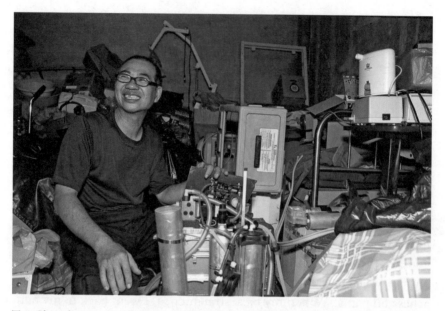

Tzu Chi volunteer Ku Si-Chiao used to have a drinking problem, but he says taking up recycling has given him a new outlook on life and helped him end his dependence on alcohol. (Photographer / Lin Yan-Huang)

reference to sociologist Pierre Bourdieu, "domain" determines a person's thinking and behavioral changes.

The field of conservation provides one of the best environments for life education, like a religious altar or temple. In fact, it is more powerful than a temple because people here have accepted the idea, not only by reading the classics to attain understanding of the concept and through listening to the Master, but through practice that will wash away their negative emotions. The Tzu Chi recycling process lets many lives wash away their long-standing bad habits.

Tzu Chi volunteer Ku Si-Chiao had a drinking problem stemming from a loss of self-esteem, for example. He became anti-social, enclosed within his own world. He was always alone, unhappy, and reluctant to communicate with others. After two decades of drinking, a chance encounter led to an invitation from some Tzu Chi volunteers at the

Xinzhuang recycling station to drive their truck. Under the guidance of a Tzu Chi volunteer, he took part in recycling work three or four days a week.

Ku did not stop drinking right away. But environmental work is sometimes hard and not particularly lucrative. He noticed that there was only a modest income from the sale of the recyclable materials. He delivered a truckload of empty cardboard cartons, and the money was barely enough to buy a few bottles of beer. This had a big impact on him: he realized that his drinking had wasted so much of his own money as well as his time. He eventually decided to devote himself to working with Tzu Chi.[13]

"Tzu Chi makes me feel that life has become very real. It is no longer an illusion," said Ku. Through activities such as meditation, home visits and environmental protection, he felt as if his heart had opened up. He gained the courage to socialize, and gradually emerged from the shadow of alcohol. He no longer drinks and his heart is full of joy and happiness.

Alcoholism, from a psychological point of view, is usually a form of escape. When a person who has low self-esteem is confronted by life's obstacles or depression, alcohol is often used to achieve a feeling of "just compensation." What is it that made Ku Si-Chiao stay away from the alcohol that had imprisoned him physically and mentally for the past twenty years? His focus on environmental protection gave him a positive view of his valuable work and that moved him away from his dependence on alcohol.

Bourdieu's habitus theory mentions that the body attains habits through the environment and its activities; the environment can change a person's temperament.[14] This change in temperament will open his world and the social structure to which he belongs. This is true of the

---

[13] Wen-si, "Walk out the Shadow of a Slap: Awakening of Wu Xi-jiao," *Tzu Chi Monthly,* 373 (1997): 66.

[14] Pierre Bourdieu, *Outline of a Theory of Practice* (Cambridge and New York: Cambridge University Press, 1977), 86.

friendly atmosphere and interpersonal relationships that can change a person's habits and temperament in an emotional dimension.

Bourdieu also describes habitus as a "structural mechanism." Its operation works from a person's own awakening, and it changes its own habits through inner spiritual awakening. This seems to be precisely what many Tzu Chi environmental volunteers are able to do; they improve themselves through the culture and value system that leads to their transformation. This theory applies to people like Ku Si-Chiao, who has been transformed both physically and mentally and has been "reborn."

Eric Fromm once said: "The blows of life on the one hand let people grow up, while, on the other hand, they may deeply affect a person's self-confidence."[15]

Environmental protection volunteer Lin Chien does not look like a gambler. Her early years were ill-fated: her husband was paralyzed after a car accident, and he passed away after being bedridden for 18 years. Her eldest daughter also had an accident and passed away after her husband died. These misfortunes left Lin Chien in extreme grief. To ease her pain, she turned to gambling. At first, she saw gambling as a distraction and bet only small amounts. Later she was betting thousands of dollars on the lottery. She was trapped. Despite pleas from friends and relatives, she continued. Her daily mood depended on her wins and losses and this affected her physically.

One day, she lost all the money she had with her. In frustration and anger, she ran home to get more money to win back her losses. This coincided with a visit by a Tzu Chi member who had come to her house to accept a donation from her daughter-in-law. At that time, Lin thought that Tzu Chi was a temple, and that the temple was raising money to build more temples. Her instinct was to donate to the temple to obtain the protection of divine providence. In fact, this brought in the "living Buddha" who changed her life.

---

[15] E. Fromm, *The Art of Loving*, trans. Meng Xiangsen (New York: Harper & Brothers, 1956; Taipei: Zhiwen Publishing Co., 1986), 25.

Lin made a donation of $500 to Tzu Chi and then hurried out to gamble again. Once again, she lost her money. After a month, the Tzu Chi member visited her house again. Lin became angry as her luck at gambling had not improved. But this did not discourage the volunteer. He brought Master Cheng Yen's videotapes to show Lin the next day. She patiently watched the videos and saw the compassion of Master Cheng Yen for the suffering of the world.

Lin later joined a tour of Hualien Tzu Chi and listened to Master Cheng Yen's speeches. The Dharma Master said: "Turn garbage into gold, gold into love, love into a clear stream, the stream will go around the world ..."[16] Lin thought about Master Cheng Yen's speech carefully and concluded there were new possibilities for her. "Every one of us has a pair of hands. We have the means to do something meaningful, using environmental protection to do good deeds. We can also learn to cast aside our reservations," she said. The Dharma Master's speech awakened her: these hands could do good deeds instead of gambling.[17]

From the perspective of psychoanalysis, how can we explain the power of gambling?

Paul Bellringer, the noted British expert on gambling-related social problems, explains the gambler's psychology: "For gamblers, gambling can be fulfilling. It is an escape from pressure. It gives a feeling of being in control and winning increases a sense of accomplishment."[18]

When Lin's husband and daughter passed away, she could not bear the enormous sorrow. She felt powerless and used gambling as an

---

[16] Shih, De-fan, *Cheng Yen Shangren de Nalu Zuji*, Vol. Winter (Taipei: Tzu Chi Cultural Publishing Co, 2006), 70.

[17] Zhang Shun-yan and Li Shu-hui, "Tzu Chi first Recycling Volunteer Convention: From environmental protection to family and mind reservation," *Tzu Chi Daolu*, 244 (1996): 2.

[18] P. Bellringer, *Understanding Problem Gamblers*, trans. Guang, Meifang (London: Free Association Books, 1999; Taipei: Zhang Laoshi Cultural Publishing Co., 2002), 110–112.

escape. It was her way of trying to regain control over her life. At the same time, she hoped to receive social warmth and affection as she gambled but this hope was never fulfilled. When she joined Tzu Chi and engaged in recycling, however, her life gained direction. She obtained satisfaction from her achievement and that gave her a sense of being in control.

The significance of environmental protection and the importance of recycling for the sustainability of mankind has given Lin Chien a positive outlook on life. The care and consideration of Tzu Chi members, their religious convictions and Master Cheng Yen's "Great Love" have given her a new sense of wisdom and an ability to accept the impermanence of life. She was able to understand that "people cannot dictate the length of life but can dictate the depth of life." All the fortune and misfortune end with an idea. Hands can gamble; they can also do meaningful work. This value of life and the warmth of a family are the main reasons why Lin has stayed away from gambling since she joined the recycling program.

Physical and Mental States

Lin and other volunteers who are ailing physically or have psychological difficulties have experienced significant benefits from joining the recycling program. This change is not entirely due to recycling work; it also reflects the effects of the love and compassion expressed at the recycling stations.

Master Cheng Yen's concept of "treasuring the life of things" was promoted by a group effort at each recycling station and reinforced by Tzu Chi's media arm, Da Ai TV. The shared experience of volunteers deeply penetrates every environmental volunteer's heart. This becomes the key driver of their physical and psychological transformation. Kurt Koffka, the co-founder of Gestalt psychology, has described similar situations, saying the sense of value connects man's inner self to the outside world; people's "psychological field" and "physical field" interact with each other mainly because the sense of their value and meaning are established:

> The mental state of the person is not a series of experiences accumulated, or direct physiological responses to the environment; it is the continuity of the significance of cognition and construction. The concept of value: helping people to understand the world and to understand themselves.[19]

The sense of meaning and value is an important tool for people to rebuild the mind. As Koffka said: "We collect, classify and acknowledge living experience, in order to help our mind establish a simple and constructive meaning of value."

Koffka further pointed out that the reward may not necessarily result from a person's ability to learn. Many became models of success through the establishment of the sense of value and significance. The physical and mental state of Tzu Chi's environmental volunteers were reconstructed and revived, not through material reward, but through the establishment of values and meaning. We observe that, in Tzu Chi's recycling centers there are many senior citizens, people with disabilities, and people with psychological or family problems. They rediscover their value of life through recycling.

Rebuilding Family Relationships

The concept of values is the key to uniting a family. Although many families have love, their concept of values differs. They have difficulty in getting along with each other and may even become enemies. At Tzu Chi's recycling stations, families are able to rediscover their common values. Through recycling, they find common themes and

---

[19] K. Koffka, *Principle of Gestalt Psychology*, trans. Li, Wei (London: Lund Humphries, 1935; Taipei: Zhao-ming, 2000), 65–128.

Recycling as a family activity. (Photographer / Huang Shi-Ze)

interests. Through the interaction of Tzu Chi's members at the recycling station, the originally cold and antagonistic family relationship is improved. A study conducted at a recycling station revealed that many families facing problems were able to mend and restore their warm family relationships.

The biggest crisis of the modern family stems from job specialization. In industrial capitalism, everyone seems to live in a compartmentalized society. Individuals not only become isolated from society, even the family drifts further and further apart due to a lack of common topics for discussion about life issues and values. The rising divorce rate is the inevitable result of this kind of society. In a materialistic environment, man faces pressure from daily competition. His desires need to be relieved through a variety of escape and comfort mechanisms. Human nature, in its demand for true love, is dampened in a highly competitive society. This repression cannot be recovered through the family, as it suppresses the gentle family unit and its mutual love.

Ulrich Beck, a German sociologist, and his wife Elisabeth Beck-Gernsheim co-authored the book *The Normal Chaos of Love* which describes this phenomenon in the civilized world. It focused on the division of labor in an industrial society and its impact on the family:

> If the social situation is forcing individuals to focus on their own interests, how much does it take to share a personal life? Even if the motive is completely perfect, the following is also bound to be the result: two world bodies that cannot build a common life, must each defend their own world and will ultimately lead to a violent dispute in this civilization, and will sometimes get out of control.[20]

Recycling stations provide a married couple with a common mission and values. The recycling station, as a big family, helps many couples who are on the verge of separation re-establish harmonious family relationships.

The family of Yang Wang Yi, a volunteer at the Grand Bay recycling station in Tainan, is an example of how this works. Her husband, a taxi driver, was considered hard working, but because of the nature of his job, he was often away from home. He acquired many bad habits — he smoked, chewed betel nuts, and had a quick temper. Family friction ensued.

Yang Wang Yi had a small grocery store and a Tzu Chi member asked her to donate the empty cans, bottles and other recyclables from her business. She knew about Master Cheng Yen's efforts to protect the environment and use the proceeds from recycling to purify the mind and provide relief to the poor. She did not know at the time that her kindness eventually would solve the family problem that had haunted her for years.

Initially, her husband strongly opposed her recycling work, voicing concerns about sanitary conditions. But, when he saw how neatly she

---

[20] Ulrich Beck and Elisabeth Beck-Gernsheim, *The Normal Chaos of Love* (Malden MA: Blackwell, 1995), 92.

organized the recycling materials and the pleasure she obtained from this work, he gradually took an interest in recycling. Her husband eventually absorbed the ways of the Tzu Chi members and gave up his bad habits. The joy of recycling work has replaced the pleasure he obtained from those habits in the past.

Tzu Chi recycling stations provide couples with a space for family activities. This domain becomes a place where they can share a common topic for communication. This domain is not competitive; instead it is creative. It does not exploit but gives individuals a sense of achievement and respect. Rather than having mutual suspicion, there is mutual respect and mutual love. This relationship provides the modern world with another kind of reflection and healing. This domain of love of nature and love of each other provides a common value system, which is the instrument Tzu Chi volunteers use in rebuilding family relationships.

Classification of Goods and the Reorganization of One's Soul

Social psychologists suggest that, if a person is unable to rely on his will to change his mood, he can gradually change his behavior by changing his temperament. Therapists in psychology use the "letting one foot in first" theory (A foot-in-the-door principle) to guide people to start with a small step, so that people with psychological distress can change gradually and broaden out into every part of their lives. From then on, a person can change his troubled body and mind and his unstable temperament.[21] The outside world to a certain extent affects a person's inner self. This is what we have mentioned previously in Bourdieu's theory that, through contact with the outside world, a person's temperament can change.

---

[21] J. L. Freedman and S. C. Fraser, "The foot-in-door technique," *Journal of Personality & Social Psychology*, 4 (1966), 195–202.

Social psychologists maintain that, if people tend to be lonely, depressed, withdrawn, and shy, it is often due to a lack of external social skills and not a result of some internal problems. In a safe place, people learn new social skills, and are able to rebuild their lives into a life of confidence. Perhaps this can be used to illustrate why many volunteers gradually leave depression behind after joining the recycling programs and overcome chronic psychological problems.

When the environmental volunteer separates items like paper, plastic and cans, his inner self seems to be going through a reclassification too. Environmental volunteers concentrate on recycling work with a pure and innocent heart. This process has a healing effect on mental ailments such as anxiety and mood changes.

Chai Man was suffering from despair and anxiety. She said her life was full of drama. Her mother died when she was four; her father passed away when she was 14. During her most trying times, at 24, all her valuables were stolen by thieves who broke into her home. A decade later, the illegal shelter she lived in was destroyed in a fire. Eventually, she slipped into deep depression.[22]

But recycling work at Tzu Chi helped her concentrate on something beyond her own problems. She focused on separating different types of plastic bottles and crushing aluminum cans. Gradually, the pain of depression vanished. Recycling was the prescription that cured her sadness — without the use of medication.

Aging and Community Cohesion

Many of those who join the environmental program are elderly and have long been living alone. To persuade such people to engage in recycling, volunteers go from house to house to encourage participation. With the passage of time, volunteers are able to share what they

---

[22] Wang Shu-fang, "The Experience of Chai-man: Seven Years Trapped in a Mind Cocoon." *Tzu Chi Monthly*, 472 (2006), 21.

have learned at Tzu Chi, such as cherishing life and the spirit of pre-serving the environment. Initially, many elderly people may be reluc-tant to take this first step. But over time, they become interested in acting as a guardian of their community and working for the better-ment of Mother Earth.

Taiwan is an aging society, but the government and private organi-zations rarely propose specific methods to address the loss of social status for the elderly. The elderly need increasing levels of care and have other problems associated with advanced age. Many senior citi-zens feel worthless. They stay at home all day without anyone to talk to. Their loneliness often leads to more self-isolation. But joining in environmental protection work gradually transforms them into outgo-ing individuals who win praise for their selflessness. They can retrieve their self-esteem through recycling.

There are thousands of senior citizens in Taiwan who are volun-teers at the recycling stations. All have experienced the transition from emptiness to fulfilment and from self-isolation to re-integration into the community. Some who hardly said a word for an entire day have become community leaders who persuade their friends and neighbors to join in recycling. Their transformation confirms research conducted by psychologists that behavioral change transforms attitudes.

The Meaning of Life

Tzu Chi's recycling volunteers often use the recycling station as a venue for meditation. As noted earlier, Master Cheng Yen always says that the recycling station is like a temple. Environmental protection volun-teers say a prayer before they start the day undertaking recycling activi-ties. Facing a blank wall, they silently pray that there will be no disasters in the world, that people's hearts will be purified and that there will be social harmony. Buddha is in everyone's mind; they do not have to dedicate themselves to a specific image.

The bonding of the environmental volunteers is a centripetal force. It is like the spirit of a religion that absorbs. This kind of religion does not seek eternal life but uses the tangible life to touch on the material life, to cherish the earth like cherishing one's body. This feeling of human, heaven and earth communicating with each other is both practical and transcendental. This is the religious outlook of Master Cheng Yen, a feasible and practical way to let people embody self-awareness and the immortal wisdom of life.

The recycling efforts to maintain our sustainable living environment reflect a kind of "intergenerational care and justice." Volunteers who participate in environmental work have a hidden psychological concern more like a religious ritual that is beyond this world, which is to pray for the endless extension of the life of the world.

Chen Ching-Wan, a volunteer from Taipei, describes this as "picking up garbage and putting down the garbage in one's heart." He contends that recycling not only takes him to a calm and peaceful realm but can also benefit others around him. While picking up tangible garbage, we also clean up the invisible garbage in our hearts, he says.

The *Sutra of Innumerable Meanings* says: "One law can take all the laws." Every individual is a small unit which can "praise the genesis of Heaven and Earth."[23] Each force is as small as a grain of sand but taken together these grains of sand can form a tower. This can be shaped into a society of true goodness and humanity. Every PET bottle that is picked up can let the world journey in the direction of continuation. Every piece of paper recycled is a small step toward the purification of the human soul. Every grain of sand is a world, and a paradise can be seen in a wildflower. Grasp infinity in the palm of your hand and place eternity into a moment of time. Nothing is insignificant for the environmental volunteers; their actions reflect the true meaning of this word. Through Tzu Chi's recycling efforts, each tiny life of each individual extends to the purification of people around the world.

---

[23] Master Cheng Yen, *The Sutra of Innumerable Meanings*, Sutra Collection-7 (Taipei: Tzu Chi Cultural Publishing Co, 2001), 182.

Chapter 3

Recycling in Action

"We need to take protecting the environment seriously"
— Master Cheng Yen

The Tzu Chi Foundation was set up in Taiwan in 1966 and is now the world's biggest Buddhist charity. It has over 100,000 dues-paying members in Taiwan. If we count those who make occasional contributions to the charity, it has millions of supporters in Taiwan alone and as many as 10 million worldwide. Buddhism is often thought of as focused on attaining individual spiritual enlightenment and perfection. While various Buddhist groups show many individual acts of kindness and compassion, the overall objective is generally seen as the transformation of the individual spirit. Tzu Chi, however, is an example of humanistic Buddhism that is fully engaged with the rest of mankind. Its very name could be translated as "compassionate relief." It applies Buddhist values in its interaction with the world at large. Its mantra, if you will, is that it is dedicated to better social and community services, education, medical care and promoting humanistic values. The fact that it embraced recycling as part of a mission to "Save the Earth" is hardly out of character.

Tzu Chi advocates simplicity and the avoidance of waste. It endorses what it calls the 5Rs — refuse, reduce, reuse, repair and recycle — or refuse to buy unnecessary items, reduce resource

41

Volunteers help load used cardboard onto a recycling truck as part of the night-time collection program. This program has helped increase community participation in recycling, allowing people to dispose of their household clutter after the workday. (Photographer / Wu Yi-Min)

consumption, reuse products, repair broken items and recycle those that can no longer be used. The aim is to make do with less and reduce the heavy human footprint on our planet. It suggests we eat less, for example. We should be able to make do with 80% of what we would normally consume and save 20% for charitable donations, according to Master Cheng Yen. Tzu Chi volunteers carry their own cups, bowls and chopsticks when traveling so they can reduce the demand for throwaway eating utensils. They bring their own reusable bags and are encouraged to walk or ride a bicycle when possible instead of driving a car. And they are called on to use less air conditioning to ease the strain on our planet's resources. The attitude of volunteer Yu Hsueh-fang is typical: "Dharma Master Cheng Yen said she uses only one washbowl of water every day. We have to learn how she cherishes the

Earth." Ms. Yu says that when she rinses vegetables as she prepares meals in her home, she saves the water to use on her plants and flush the toilet.

When Tzu Chi started out on its recycling campaign it had to overcome a deeply ingrained view that sifting through garbage to find recyclable material was an unpleasant task performed only by those who had no other means of earning a living. Master Cheng Yen asserts this was in fact an effective way to promote humility. The persistence and tireless efforts of Master Cheng Yen and her followers overcame these challenges and turned this into an international mission that inspired people around the world. From toddlers to centenarians, from farmers to doctoral students, from the physically challenged to ambassadors and titans of industry, all have joined in to show their love for the community and the planet we share. And in this task, they experience the value of life.

No shortage of paperwork. Recyclers sort paper at a Taiwan recycling point. (Photographer / Zhang Mu-Lan)

In its early years Tzu Chi's recycling campaign was little more than a collection of individual actions. Yang Shun-ling was among the first to respond to Master Cheng Yen's call to action. She was in the audience when Master Cheng Yen made her first public remarks on the need to recycle. Ms. Yang began to recycle waste in her daily life and donated all the money she received from the sale of the recyclables to charity. Tsai Kuan, now 101, was another early volunteer. She too was on hand when that first call to action was made. She relates how she would get up at 4 am to start her rounds in the search for recyclable material. She says her efforts were not universally understood at first, but through persistence she managed to convince others of the value of recycling. Another early volunteer was Tseng Ching-hui. "We had lots of obstacles initially," he says. He used a space in front of his home to sort recyclable material. Word got out that he was collecting discarded materials, and in the evening, people would drop off all sorts of unwanted items. He ended up with an unsightly collection of old bathtubs, refrigerators, and other appliances. Neighbors complained of the piles of cast-offs and even the Environmental Protection Administration came around and issued fines. "We were accused of 'making merit' at the expense of our neighbors," he says, referring to the Buddhist ethos of accumulating virtue through positive actions. "We had to overcome a lot of obstacles."

One of Tzu Chi's first recycling centers was in the village of Liming in Taichung, not far from the site of the lecture that kicked off the recycling effort three decades ago. The site was donated by Chen Juanchai, a local volunteer who said she wanted to keep her community as clean as her own home. More recycling centers followed, and the campaign quickly moved overseas. By 1991, Tzu Chi volunteers were using the home of a volunteer in the Los Angeles area for recycling and before long Australia had a recycling campaign too. By 2007, Tzu Chi had expanded to 5,235 permanent recycling centers and temporary stations with 70,950 volunteers in 14 countries and regions. And in 2019 there were 532 permanent recycling centers, 9,622 temporary recycling stations and 112,016 volunteers in 19 countries and

Recycling operations in Malaysia. (Photographer / Luo Jin-Xing)

regions. In Taiwan alone there were 89,585 volunteers and 8,419 permanent and temporary stations.

Turning Garbage into Gold

Some of the recycling stations have had a surprising history. In Mucha on the outskirts of Taipei, for example, a recycling station has sprung up on a site that once housed the clandestine manufacturing operation of a criminal drug ring. It's out of the way, perhaps explaining its choice as a center of illegal activity in years past. The drugs are gone now, replaced by stacks of cardboard, paper, plastic bags and discarded home appliances. Trucks bring in the raw recyclable materials and then haul them out for delivery to traders or factories once they have been cleaned and sorted.

On a recent visit, the workday was just getting started but the volunteers — about 30 of them — were already busily sifting through mounds of unwanted materials. They were sorting plastic bottles, cutting off the plastic rings from the bottle tops, and separating the colored containers from the clear ones. This is what the volunteers describe as "turning garbage into gold." By separating the types of plastic, they increase the purity of the materials and in turn boost the value. The recycling volunteers start arriving at about 6 am; some coming from considerable distances away. Many of them are old enough to be enjoying their retirement years, relaxing at home, or keeping an eye on the grandkids. Instead, they feel they have a mission and willingly contribute their time and energy to a good cause. They work in less than ideal conditions through midday or longer. It's a dirty task and without pay, but there is the camaraderie of like-minded recyclers. They know they are making a contribution to something bigger than themselves — easing the stress placed on our environment. And there is something else they can look forward to: a hearty lunch afterwards.

"They come here without any pay," says Kao Pao-tan, who is in charge of the operation. "It's hot in the summer and cold in the winter," she says, noting this doesn't dent the enthusiasm. Ms. Kao, 65, has devoted the better part of the last decade to recycling work. "Sometimes I drive the recycling trucks," she says with pride.

One of the regular volunteers is 52-year-old Hu Su-chen, who has been working since she suffered a stroke several years ago. She says the regimen is in fact a useful form of physical therapy. Exercising her fingers brings her closer and closer to full recovery. "The more I work, the healthier I am." And it has another benefit: "It makes me happy," she says.

That sentiment is shared by 73-year-old Lu An-tien, a former engineering manager at what was then known as General Instrument Corporation. He sits beside stacks of old fans, CD players, LED lights, motorcycle helmets, vacuum cleaners, thermos bottles, and PC boards. A new-looking robotic floor cleaner has been stowed near a beat-up laptop computer, some dusty equipment from a dental practice and a long-forgotten water dispenser from someone's office. Mr. Lu's job is to repair these cast-offs, if possible. If not, he dismantles them and saves anything valuable for resale.

"I'm extending the life of these products," he says, cheerfully referring to a key goal of the recycling mission. This work reminds him of what he did in the early years of his career. "I did things like this 30 years ago. My job then was to take things apart."

He says he got started in recycling with an introduction from a neighbor, a refrain echoed by many volunteers. One person in a neighborhood takes an interest in the effort to recycle and reduce waste. The next thing you know friends and relatives are eager to join in. The recyclers have a clear goal: protecting the environment for the benefit of future generations. "We need to do this on behalf of our next generation. We have to give them a clean place to live," says Mr. Lu.

Collecting Trash in Style

Chen Chin-hai was one of the first members to join in the recycling drive in Luzhou in New Taipei City, which surrounds the capital Taipei. He and his wife were running a successful construction business when they took an interest in recycling.

"We started recycling in 1992 with a handful of Tzu Chi members," he says. "The volume of garbage was growing very fast and we got government people and retirees to participate. We too are turning garbage into gold."

Eventually, Mr. Chen turned the day-to-day operations of his business over to his partners so he could concentrate on recycling. "Now I am here full time," he says as he surveys the Luzhou recycling center.

In the early days he drove around the neighborhood picking up all sorts of waste materials in his family car — a Mercedes Benz. "I may have been the only person driving a Mercedes to collect garbage," he says.

In 2003, a night recycling program was started up, allowing people the chance to dispose of household clutter after the end of the workday. Once a week, teams of recyclers would set up temporary collection points at locations around Luzhou. Neighborhood residents would drop their recyclables off at one of the temporary collection points.

Businessman Chen Chin-hai was one of the first to join the recycling cause in Luzhou near Taipei. A successful businessman, he turned his company's day-to-day operations over to his partners to devote himself to recycling. Tzu Chi volunteers often speak of turning garbage into gold. The sign behind Mr. Chen reads: "Turn gold into love." (Photographer / Huang Shi-Ze)

Once the recyclers had enough of a haul, small pickup trucks driven by volunteers would cart the night's collection to the neighborhood's permanent recycling center.

Driving around one evening while the weekly collection was under way, Mr. Chen checked in with groups of recyclers gathered at the entrance to local parks, in front of apartment blocks or in cramped doorways. The recycling points ranged from the busy main streets to quiet back alleys. About 200 volunteer recyclers, often called "Recycling Bodhisattvas" — referring to the Buddhist term for someone on the path to enlightenment — took part in the neighborhood collection drive.

"We do this every week, all year-round except for Chinese New Year," calls out one enthusiastic recycler as Mr. Chen stops by for a

A volunteer snips a hard plastic ring from an empty plastic bottle. Separating types of plastic increases the purity and the value of the recyclable materials. (Photographer / Ye, Jin-Hong)

quick inspection of the collection drive. Satisfied with what he sees, Mr. Chen hops back in his car and heads off to another collection point.

While recycling is a mission that can motivate thousands of dedicated people, it is not always so easy to tolerate mounds of discarded items suddenly appearing near one's front door. Noble goals are fine in theory but a "not in my backyard" sentiment often prevails. Tzu Chi has worked hard to make sure the recycling process is as trouble-free as possible. It engages with residents and government organizations, seeking their approval to use nearby public or private spaces for a few hours a week. Volunteers spread out a canvas to cover the ground and keep the mess to a minimum. Piles of unloved household items — shoes, luggage, cardboard boxes — now seen as no longer useful, start appearing as residents arrive, lugging their soon-to-be abandoned possessions. The objective of the donors is twofold: clean up their household clutter and perhaps find some use for these cast-offs with the help of the caring hands of the Tzu Chi recyclers.

The Recycling Bodhisattvas are volunteers whose main reward is personal satisfaction. They quickly sort the items and toss them onto open trucks; many of them driven by owners of businesses in the neighborhood who similarly feel the need to participate. When the work is done, they hose down the recycling area so that the dirty job of collecting trash leaves little or no visible trace.

At one collection point in front of the local post office, three-and-a-half year-old He You-chin is helping her grandma sort plastic bottles that are piled up on a canvas spread out over the pavement. Grandma Roan Li-huai, who says she has been a volunteer for six years, gives helpful instructions to the young recycler on where to place different types of plastic bottles.

The materials collected in neighborhood temporary stations are gathered up and sent to the local permanent recycling center. Here there are more signs of recycling as a family affair. In one corner a family of four is busily sorting the contents taken from one of the pickup trucks while a father and son team unload another truck nearby. On the front wall of the permanent recycling center is a sign with a reminder of the guiding principle for Tzu Chi's army of volunteers: the 5Rs. The sign is helpfully in Chinese as well as English.

The money from the recycling operations goes to fund Da Ai TV, Tzu Chi's television station, which blends a message of the teachings of the Buddhist group's founder, including the need for recycling and consuming less, with entertainment and reports on Tzu Chi activities. It calls on its followers to "Cherish the Earth" — a message that complements the call to recycling. It is a theme that is reinforced in homes and schools. Young children repeat this golden rule, sometimes reminding forgetful elders.

Fortunately, Tzu Chi is not operating in a vacuum. The government has been steadily adding to its array of regulations for household and industrial recycling. The Solid Waste Disposal Act of 1988 made producers and retailers responsible for disposing packaging containers. In 1991 the bottle deposit system was set up whereby each bottle purchased included a refundable NT$2 payment, and later this was shifted to a subsidy to businesses in the trade. It was highly successful at getting people to participate, and even after the incentive was removed, recycling continued uninterrupted as a way of life. (The program was perhaps a little too successful at one point — the recycling rate soared to 120% of total reported PET bottles in Taiwan. This was due to a surge in imports of used plastic PET bottles in an effort to take advantage of the subsidy.) The subsidy was reduced and ultimately abolished in 2002. By 2018, long after the subsidy had been scrapped, the recycling rate for plastic PET bottles stood at an enviable 95%.

In 2005, mandatory sorting was rolled out and in 2011, the first environmental education act took effect. Executives of all companies of a certain size were required to take courses related to environmental issues.

The efforts of groups like Tzu Chi coupled with the government drive to promote recycling has produced an odd measure of success for the Buddhist organization; the volume of materials recycled by Tzu Chi, both as a percentage of all recycled materials and in absolute terms, has actually fallen in recent years. As others pitch in, the role of Tzu Chi — while still important — has declined. Tzu Chi once accounted for as much as 30% of all recycling in Taiwan, according to government statistics. But Tzu Chi's activism, combined with strong government support for environmental protection, has helped promote recycling in the public eye. Now Tzu Chi accounts for only about 3% of the total. In absolute terms, the total volume of materials recycled by Tzu Chi stood at 81,831 metric tons in 2019, down from a peak of 152,362 tons in 2006, though still well above the 1,710 tons recorded in 1992.

"We now have less trash to recycle," says Mr. Chen, noting that the government and private businesses are accounting for more and more of the total. "This is actually a sign of our success."

The Fruits of Our Labor

Changhua, in central Taiwan, has a number of claims to fame. One of them is a huge statue of the Buddha at Baguashan, which looks down benevolently on the city below. The other is that it is one of Taiwan's main guava-growing areas. (In 2019, the U.S. and Taiwan signed an agreement allowing imports of fresh guava from Taiwan into the U.S. market, making Taiwan the first approved Asian supplier of the fruit for American consumers.)

So, perhaps it should come as no surprise that a Buddhist organization would use the humble guava as a starting point for recycling in Changhua. Fresh guavas are brought to market with a protective web-like plastic cover. While this helps shield the fruit from bruising before it reaches consumers, it creates a significant waste problem with a considerable impact on the environment.

"There's a huge volume of polypropylene wrappings here. If we didn't do this you could imagine the impact," says Shi Shu-yin, a volunteer at a collection point in Yuanlin in Changhua county. The volunteers sell the plastic to factories that can reuse the material to make new plastic protective wrappings. It's a lot of work for a little bit of financial return — at present about NT$2 to NT$4 per kilogram.

"We have a sense of accomplishment and you can't put a price on that," says Ms. Shi. Painstakingly, she removes guava plant leaves from the web-like plastic covers. "When we scrounge through the garbage for these we have to clean and separate the plastic from old socks and other things — even clean off dog feces. Everything. You name it."

Some critics have complained that this takes away a livelihood from people who rely on collecting these discarded resources for a living. Tzu Chi says it is aware of the potential conflict of interest and tries to avoid competing with private recyclers who depend on income from the sale of these materials. In most instances, Tzu Chi does the work that ordinary businesses — which are focused on turning a profit — wouldn't bother to do. Without Tzu Chi taking on this kind of work with low financial returns, this waste would end up in

Tzu Chi volunteer Tsai Kuan is 102 years young and an active recycler. She was among the first volunteers to answer Master Cheng Yen's call to join the recycling campaign. (Photo provided by Tzu Chi Foundation)

an incinerator, according to volunteers. "We don't want to add to the air pollution here," says Ms. Shi. Air pollution is already a sensitive issue in Taiwan, particularly in this central part of the island which has a big coal-fired power plant on its doorstep. Air pollution has been at the heart of local election campaigns in recent years. Many residents of central Taiwan complain they are being asked to promote progress elsewhere on the island while they pay the environmental cost.

Liu Hsueh-sheng, another member of the local recycling team, says his home is next door to a big guava market. "I can fill up two truckloads (of guava wrappings) a week." Mr. Liu says he got his start in recycling six years ago and the results have paid off. "If we didn't do this the environmental impact would be huge. All of this would go into the incinerator."

Tzu Chi's entry into the effort to clean up the environment predates the latest political controversy over energy and air pollution, but it stems from a similar motivation. "We really wanted to avoid incinerating these bags," says Mr. Liu.

One of the volunteers in this effort is 101-year-old Tsai Kuan. "I attended that lecture by Master Cheng Yen," she says, referring to the 1990 address that propelled Tzu Chi on its recycling mission. "That's when I got started in recycling. I would put things aside in the hallway. I had a little cart and I would ride around the neighborhood. People would laugh and ask how can your family let you do this? This was because they looked down on this kind of work."

"I would say I am donating the money and saving the Earth," she says. "I'd collect scissors, paper, old pieces of rope. I'd get up at 4 am to start work and go through the garbage." She too acknowledges the criticism from some quarters about depriving some low-income people with a means of making a living but echoes the theme that Tzu Chi is careful to avoid such competition. "We don't compete with anyone," she says. "If Tzu Chi doesn't do it, no one will."

She is spry and still starts her day early in the morning. She spends her time volunteering and listening to lectures from Master Cheng Yen. She moves easily and says that the exercise has helped keep her looking much younger than her years. "This has great medicinal value," she says.

Money is Not the Objective

So how much have Tzu Chi's recycling operations actually accomplished? In 2019 alone, Tzu Chi recycled 3.8 million kg of PET bottles, 3.7 million kg of plastic bags, 11.2 million kg of glass bottles, 903,437 kg of aluminum cans, 294,832 kg of copper, 187,000 kg of batteries and 39.55 million kg of paper. Since 1995 it has recycled a total of 1.439 billion kg of paper, saving 28.79 million 20-year-old trees (50 kg of recycled paper is calculated as equal to one 20-year-old tree).

Tzu Chi recycled 110,052,263 kg of PET bottles between 1996 and 2019. Some of that total went into the production of the more than 1 million blankets distributed to refugees, disaster victims and other people in need.

But the financial returns have not always been steady.

In early 2018, China banned imports of 24 types of waste materials, eliminating demand from a major market. That dealt a sharp blow to the global recycling business as waste previously shipped to China suddenly became available at cut-rate prices. The price of paper for recycling use slumped from more than NT$4–NT$5 per kg to less than NT$2. Taiwan's paper makers, like their counterparts in Southeast Asia, took advantage of the decline in international prices for paper and other raw materials and switched to imports. Taiwan's imports of wastepaper climbed to 1.34 million metric tons in 2018 from 1.1 million tons in 2017 while imports of scrap plastic waste surged to over 429,000 metric tons from 202,000 tons. That posed new problems to Tzu Chi's recycling operations. And it wasn't the first time that Tzu Chi and other recyclers had been confronted with slumping prices that would have deterred profit-driven companies.

In 1997, following the Asian financial crisis, prices for waste materials tumbled as demand declined. Traders were reluctant to take recycling materials, but Master Cheng Yen insisted that even with no return Tzu Chi would persist. "We are doing this for the Earth," she said. Tzu Chi's recyclers in Malaysia reported similar problems during the economic downturn. Tzu Chi's operations in that country ended up giving away some of its recycled material, which had lost much of its value, until an economic recovery revived prices and created fresh demand.

But Tzu Chi has been cushioned somewhat against the full impact of price declines and an industry-wide trend of narrowing profits. In recent years, buyers of waste materials used in recycling have insisted on greater purity of the raw materials. Tzu Chi had already gained a reputation for maintaining high levels of purity by carefully separating different types of plastic. That made Tzu Chi a desirable supplier even as demand weakened. That has helped the Buddhist organization fetch relatively higher prices for its recyclables and earn more income for use in other good works.

Chapter 4

Health in Body and Mind

"Extending the Lifespan"
— Master Cheng Yen

Huang Hsu Chiu-ying was badly injured in a car accident years ago. She still struggles to express herself as the injury has impaired her speech. But with persistence she makes her point, saying that thanks to Tzu Chi she got a new lease on life and a chance to participate in recycling.

Standing next to a makeshift shrine that has been installed alongside piles of used paper and worn-out household goods, she says she once thought she'd be in a wheelchair for life. But treatment at a Tzu Chi hospital proved effective and soon she was able to walk with the help of a cane. Now she walks easily. "I'm really happy to be able to do this. I got a new lease on life here," she said, adding that she has been a volunteer for more than seven years and is now a Tzu Chi commissioner.

After taking two buses to get to the Sanchong recycling center, she ties up stacks of newspapers. When she is done she turns her attention to a pile of plastic bottles, stepping on them one by one to crush them and save space before they are carted off to a factory where they are turned into plastic raw material. She takes off the bottle caps, placing the hard plastic in a separate bin from the softer bottles themselves. "This is my rehabilitation therapy," she says. She quotes one of the mottoes of the recyclers: "Extend the lifespan of things" and suggests this applies to herself as well.

Hsiao Chiu-ying is another volunteer who enjoys feeling like she is part of a bigger cause. She is 65 and has been a volunteer for the past three years. Her vision is severely impaired, but she makes the effort to come to the recycling center by bus — and mostly on her own. She has a sign that indicates which bus she wants to take to get to the Sanchong recycling center. People at the bus stop help steer her to the right bus. Once she is on board, a recorded announcement tells passengers which stop the bus is approaching, so she knows when to get off. Someone from Tzu Chi meets her at the bus stop and takes her for the short walk to the recycling center.

She soon joins a small corps of volunteers, sitting on a low chair beside a pile of books and magazines that are being prepared for their new role in recycling. Pages are ripped from their binding and tossed in a bin as the first step in the recycling process. Next on the scrap heap, awaiting its unceremonious turn, is a Chinese language translation of *Harry Potter and the Half-Blood Prince.* Hogwarts and all of its witchcraft and wizardry are no match for the steady hands of Ms. Hsiao and her fellow recyclers. Soon Harry, Professor Dumbledore and evil Lord Voldemort are consigned to the bin, taking their place in a jumble of loose pages from a legal textbook entitled *Introduction to Civil Law* and a magazine on potted plants.

"This keeps me active," says Ms. Hsiao as she vigorously tears out pages from a book handed to her by another volunteer. "The work makes my hands more agile."

Chiu Huang Ying Pi is another volunteer at the Sanchong recycling center. She is 92 years old and says she has been a recycling volunteer for the last five years. First her son came to the recycling center to pitch in, and eventually, she followed. "Now I come every day," she says, adding that she too finds the work therapeutic. "I used to have muscle spasms but not anymore. I also sleep better."

She has a cheery disposition and the physical dexterity of a much younger person. She rides to the recycling center every day on the back of a motorbike, according to her co-workers. Her son and daughter share the role of taking her to and from work each day. One of her fellow volunteers quotes her as saying that if she stayed home, she'd probably be spending much of her time trying to avoid being drawn into petty family squabbles. Recycling is so much better, helping her avoid stress and giving her purpose.

This is a common sentiment among volunteers, according to Hsu Mu-chu, a professor at Tzu Chi University's department of human development and dean of the Tzu Chi College of Humanities and Social Sciences.

"We see many elderly volunteers doing recycling work. There is a benefit to their body and spirit. They make a contribution to society and in so doing they find their own value," he says.

Besides contributing to a cleaner environment, in many cases it makes people gentler, says Professor Hsu, adding that it can change a person's lifestyle and make interpersonal relations smoother. "I know of one volunteer who brought his father to the recycling center. His father had been drinking for many years and had a number of health issues. But after a year of volunteer work his health improved and drinking was no longer a problem."

In addition to the effects on the individual, there may be benefits for family, friends and neighbors, according to Professor Hsu.

"Many of our female commissioners say that initially their husbands were not very supportive of their volunteer work. But in the end they were touched by what they saw at the recycling stations, and even decided to join in."

Besides contributing to protecting the environment and improving the physical and mental health of the volunteers, recycling aids the local community in other ways. If there were no recycling stations, the government or other civic groups would need to provide more programs for the elderly, he says.

Lee Jia-Fu, a doctor who is deputy chief of the community medicine department at the Tzu Chi Hospital in Taipei, makes a similar assessment. He notes that Taiwan is rapidly becoming an aging society,

Dr. Lee Jia-fu, deputy chief of the community medicine department at the Tzu Chi Taipei Hospital, speaks to a recycling volunteer about his health status. Tzu Chi has many elderly volunteers at its recycling operations and its medical personnel regularly monitor the health of its recyclers. (Photographer / Chen He-Jiao)

and the need for services for the elderly is growing. Those 65 and over are already 14% of the population and that will rise to over 20% in seven years, making Taiwan a super-aging society.

"That means dementia will be a big problem," he says. Many of the Tzu Chi volunteers are elderly so these age-related issues are important to the charity organization as well as the community at large. "Recycling is one way to slow the pace of mental deterioration. Sorting helps keep people mentally alert."

Taking part in the recycling effort with other senior citizens can have another beneficial effect as it wards off loneliness. "Many elderly people are alone and despondent. But once they come to the recycling center, they discover this task has value and they find they too have value. We have many anecdotal stories of people more at peace with

Huang Hsu Chiu-ying was severely injured in a car accident years ago but she says recycling has given her a new lease on life. (Photographer / Chen Zheng-De)

themselves," he says. "They get pleasure out of this work and they also gain respect. This is very important," he says.

"In general they feel better. They eat together, work together and in some cases go to classes together," he says. "I have met so many who say they feel much better because of Master Cheng Yen."

He says many patients who have been recovering from illness tell him they are keen to get back to recycling soon. They see illness as depriving them of the chance to participate in an important community effort.

Just to make sure that the volunteers stay healthy, medical workers make regular visits to the recycling centers. They check the blood pressure of volunteers and keep track of their general state of health. The recyclers also do light calisthenics during the day, taking a 10- or 15-minute break from sorting. Those who are younger help the older ones when they are at work and when they are doing their exercises. "Participating in recycling activities gives them a sense of caring for others," says Dr. Li.

Long-Term Care

As an aging society, Taiwan is facing a rise in age-related social and health problems as well as a growing need for elder care services. Tzu Chi's recycling centers provide some of these much-needed services, supplementing the efforts of the state.

In some of these areas, Tzu Chi works with the government to fill the gap. "Some of the training for our long-term care volunteers is done under a cooperation program with the government, which actually subsidizes some of the cost for Tzu Chi and other service organizations (such as nonprofits and for- profit private companies). We

Dr. Tsao Wen-long, head of the center for dementia care at the Tzu Chi Dalin Hospital, holds a memory enhancement class for elderly recyclers. (Photographer / Wang Cui-Yun)

provide home care services under this program," says Tseng Ya-lan, a team organizer for Tzu Chi's long-term care services in Pingtung in southern Taiwan.

"Sometimes these services are for our own volunteers who may not be able to come to the recycling center for reasons such as they have fallen and injured themselves or need to take care of a spouse or other family member," she says. "We let them know we still care for them. The government doesn't have enough manpower to do all of this, so it works with us."

The government partners with Tzu Chi because it is a big and reliable non-government organization. "Tzu Chi has a good foundation and is very detail-oriented. All of this allows us to expand the scope of our recycling program," she says.

So, what services can be provided by Tzu Chi that are beyond the reach of the government?

"One example I can think of is there is an elderly person in our care who lives alone. As the Chinese New Year approached, we had to keep track of how he was doing. This is a long holiday and few restaurants stay open at this time of year. So, we look after people like him. The government is not able to do this, but we have the human resources," says Ms. Tseng. "We also deliver blankets to people who might be in need of them. These blankets are lightweight and easy for elderly people to use."

Pingtung is an agricultural area with a high proportion of elderly residents as many younger people have moved away looking for employment opportunities elsewhere. "Since the local population is older, we have a lot of dementia issues that we need to deal with. Our aim is to have a long-term care center not too far from the people who require our services."

As a result, there are more than 10 centers in Pingtung. "The recycling centers are actually a mixed model where long-term care operations are integrated with the recycling program," says Ms. Tseng, a social worker who joined Tzu Chi in 1999.

Some of the Tzu Chi volunteers help the elderly do light exercises. In some cases, the elderly are given simple recycling tasks in an effort to ensure they get some exercise and their muscles don't atrophy. Light exercise helps with the use of their hands and helps delay the loss of flexibility in their fingers.

"If we can extend the time spent out of old age homes that is a positive result. The elderly will have more time with their families," she says. "That helps them psychologically. Lots of older people in Taiwan don't want to go to old age homes. They often see that as just waiting for death."

At the long-term care centers, much like the recycling operations, younger senior citizens take care of the older ones. The object is to make all of their lives more worthwhile.

"If seniors get respect, eat well and stay healthy, they will come out of their home and interact with people. This keeps them in a better frame of mind," she says. "In many cases we also see an obvious sign of improvement in their mental outlook with less depression. They sleep better and are more talkative. The positive results are very plain to see."

Chapter 5

A New Life — Reintegrating into the Community

"Repentance Purifies the Mind"
— Master Cheng Yen

On a Sunday morning about 30 former prisoners arrive at a Tzu Chi recycling center in Changhua in central Taiwan. The ex-prisoners are there for a brief chat with a government social worker who monitors their case. He asks about their health and job situation, checking off items on a form that gives him a handy glimpse of their progress since they were released. The Ministry of Labor helps them find work, though some have trouble holding down a job. Their social worker, Mr. Sun, runs through a list of questions. Are you employed now? How many hours a week do you work? Have you borrowed money? How do you spend your free time? Any gambling? Any religious activities? Mr. Sun then discusses their case with a small group of Tzu Chi volunteers seated nearby.

Now in its third year, this is a cooperative program between Tzu Chi and the government. The released prisoners have an opportunity to interact with Tzu Chi members as part of an effort to change their behavior and ultimately adjust their outlook on life.

Yang Chi-wei is a gengsheng ren, someone who has found a new life after serving a prison term. He once was a member of the elite Hailong or "Sea Dragon" frogmen but gambling and theft landed him in jail. Now he is reintegrating into society with a helping hand from Tzu Chi and recycling. "I led a wasteful life, but recycling has helped me change," he says. (Photographer / Zhan Da-Wei)

"We work with the government," says Chiang Feng-tsuo, who is in charge of the recycling operation where this group of ex-prisoners is asked to report each month. Mr. Chiang, working with his wife Lu Hsiu-eh, began a recycling operation in another location in Changhua more than two decades ago. Although the location has changed, Mr. Chiang's enthusiasm for recycling has not. He also sees recycling as an effective means of helping former prisoners reintegrate into society. "Tzu Chi wants to change them," he says of the ex-prisoners. "And this requires a long-term social care framework."

The former prisoners also report to a government office once a month but the atmosphere at Tzu Chi is far more relaxed. The Tzu Chi office is on a green and leafy site with a modern lecture hall for larger gatherings. Nearby there is a spacious warehouse for sorting and storing recycling materials while in a far corner there is space for repairing or dismantling discarded equipment. Today, the meeting hall has been converted into a huge dining room. It is just ahead of the Chinese New Year and Tzu Chi is about to hold its year-end "thank you" lunch for all of its volunteers. All of the ex-prisoners are their guests. Like everyone else, the former prisoners line up for a heaping plate of vegetables and rice. This is a buffet style vegetarian meal, in keeping with Tzu Chi's plant-based diet guidelines. There's plenty more food for those who are in need of another helping. Some of the ex-prisoners are here because they are required to attend as part of their parole arrangement. Others stay on after the mandatory attendance period is over, finding a sense of purpose and a welcoming community they can join.

One such person who has found a home at Tzu Chi is Chen Chun-chu. She was sentenced to life in prison for drug-related

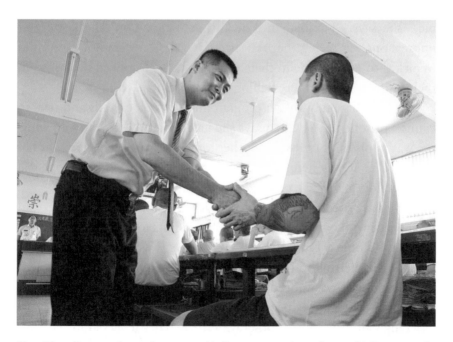

Kao Chao-liang — better known as Ah-liang — was in and out of jail as a youth. Now he's a Tzu Chi member and motivational speaker who guides other ex-prisoners towards a more fulfilling life. "I tell prisoners that change is possible," he says. (Photographer / Huang Xiao-Zhe)

offenses, but now she is known as one of the "gengsheng ren" — someone who has changed her life.

"I spent 18 years in prison," she says, adding that she has been out of jail for five years. She describes her odyssey as one that started with poor choices of friends and associates. She says she followed a long downward path of antisocial behavior. She shies away from some of the painful details, but others fill in the blanks, saying she was financing the illegal drug trade in an era when the courts handed down heavy sentences for drug offenses. As part of her parole terms, she must make regular reporting visits to the government office. "In jail I had a lot of time to reflect on my actions," she says, noting a particular regret for all the time that could have been spent with her family. "I have a big debt to my daughter," she says ruefully.

"This has been a difficult transition. There have been lots of obstacles. When I got out of prison, I had nothing, but Tzu Chi helped me," she says. "I joined in Tzu Chi activities. One of the deepest impressions was seeing that the sisters at Tzu Chi were not too proud to pick through garbage in their recycling tasks. I noticed the sisters in their uniforms. They were willing to do so much."

Tzu Chi has an outreach program: it goes to local jails and tries to counsel prisoners. Ms. Chen says she had contacts with Tzu Chi while she was in prison, but it was another former prisoner who brought her to Tzu Chi. It was here at the recycling center that she got her real introduction to a way to change her life.

"We have used this experience to get to know ourselves better and be at peace with ourselves," she says. She points to another ex-prisoner who has become a Tzu Chi recycler and says: "We were 'classmates' in Taichung," referring to their time in prison together. "Later I was sent to Kaohsiung. I was moved by what Master Cheng Yen said — there are no evil people, only people who do evil things."

Yang Chi-wei, 38, is another gengsheng ren. He served in the military, spending some of that stint as a member of the elite corps of Hailong or "Sea Dragon" frogmen. That would prove to be a proud but all-too-brief period of his life. Mr. Yang was jailed when he was 26. "I gambled and stole things. I fenced stolen goods. I led a wasteful life, but recycling has helped me change."

On release from jail, he had no money and no prospects. "I got a job doing repair work," he says. "My boss was a Tzu Chi member." That was his introduction to Tzu Chi and in the years since, he has become a regular recycling volunteer.

"I did work repairing fans and other electrical equipment. I had some contact with Tzu Chi while I was in prison. On release I became a new person. I returned the money I stole. I felt ashamed," he says.

"Now I lead a less wasteful life. I help reduce the amount of trash. This makes me stronger, and overcoming obstacles becomes much easier. I do recycling and I find I have recycled myself," he says. "Whatever I gave, I got back much more in return," he adds.

Kao Chao-liang, better known as Ah-liang, was the person who steered Mr. Yang on the path towards his new life. Mr. Kao is another former prisoner. He says he was jailed repeatedly for drug-related crimes but found the strength to change his ways. Now he is a lecturer and motivational speaker who spends much of his time talking to other youngsters who have strayed. He too is a "gengsheng ren" — having started a new life with the help of Tzu Chi.

"In primary school I didn't like to study. I later used amphetamines and I was in and out of jail," he says. "My family didn't believe I could change."

He says it was during his fifth stint in jail that he began to think of ways to break this vicious circle. He says he tried to read anything he could get his hands on, and that included Tzu Chi magazines. He had a set of earphones and would listen to Tzu Chi lectures. He quit smoking and began to realize that "we are not here just to enjoy things."

On release he created a new network of friends. "I would ride my bike to Tzu Chi training courses," he says. "My family didn't think I could make the change. A policeman told my family 'you have to be prepared for failure because many released convicts slip back into their old ways'."

But Ah-liang says he persevered and an important factor in his success was the woman who eventually became his wife. He met her at Tzu Chi. "I tell prisoners that success is getting up from the depths," he says. He adds that he corresponds with hundreds of prisoners and ex-prisoners, reinforcing the message that change is possible. "But you need to adjust their value system," he says.

Now he lectures and serves as coordinator with the government parole office. He says he is an example of "using one person to influence another." He has visited almost all of the prisons around Taiwan, he says, and at each one he brings his personal message of inspiration. "First you need to feel shame. That was what allowed me to change."

Without a Soul, Like a Zombie

Ah-liang is not the only one who has helped other prisoners on the path of finding a new life. Chang Ming-chi was a drug addict who served several terms in prison. But he found a new life without drugs, after meeting members of the Tzu Chi Foundation and becoming a volunteer himself. He now lives in Taichung, in central Taiwan, and has also become an effective motivational speaker. He too visits prisons to describe his experience to inmates, working to inspire them in the same way as Ah-liang.

"I grew up in a broken home," he was quoted as saying in an article on the official Tzu Chi website. "I felt neglected by my mother because she had to raise four children single-handedly. At the age of 17, I turned to drugs. During one six-month period, I would spend about US$300 a day. When I ran out of money, I committed robberies." For over 10 years, he was in and out of jail. "I would inject drugs. I did not know what to do. I lived an aimless life. My sister said that I was a person without a soul, like a zombie."

Chang Ming-chi is an ex-prisoner and former drug addict who now is a Tzu Chi volunteer and motivational speaker. He too visits prisons to describe his experience to inmates, working to inspire them in the same way as Ah-liang. Many of these prisoners join the ranks of Tzu Chi's recycling teams as a way to make a positive contribution to society. (Photographer / Zhang Shu-Hua)

Then he encountered volunteers from Tzu Chi who were very supportive and helped him find a job when he came out of prison. "I was able to face society because of Tzu Chi," he said, adding he has stopped taking drugs and quit smoking. He also has become a vegetarian and keen member of the foundation who adds that his volunteer work helps him atone for past sins.

Chapter 6

Vegetarianism — Better Health, Better Environment

"Following a vegetarian diet is the true way to protect the environment by practicing purity at the source"
— Master Cheng Yen

One of the tenets of Buddhism is to protect life, and that means Buddhists should avoid killing animals. Master Cheng Yen preaches this fundamental theme to the members of Tzu Chi, insisting that followers should not eat the flesh of any sentient being. But she maintains that the main reason for advocating a plant-based diet is to ensure better health and limit mankind's impact on the environment. Over the years, her views on this issue have evolved. In 2009, she had this to say about adhering to a vegetarian diet: "I used to say just try your best. But now you really have to do this ...For your health, to save the Earth, we all have a responsibility."

Research has tied red meat to increased risks of diabetes, cardiovascular disease and certain cancers. Studies have also pointed to an elevated risk of mortality from red meat intake. According to the National Institutes of Health of the United States, compared to people eating diets rich in white meat or plant-based protein, those with a diet rich in red meat had triple the levels of a chemical linked to heart disease — trimethylamine N-oxide (TMAO) — a dietary byproduct that is formed by gut bacteria during digestion. There is further evidence that limiting red meat over time can help prevent type 2 diabetes and reduce hypertension.

A study of the Tzu Chi community showed lower risks of stroke among vegetarians than non-vegetarians. In one cohort, vegetarians had lower ischemic stroke risk and in a second, they showed a lower risk of both ischemic and hemorrhagic stroke, according to Chin-Lon Lin, MD, of Tzu Chi University in Hualien, and his co-authors of the study.

"Overall, our study found that a vegetarian diet was beneficial and reduced the risk of ischemic stroke, even after adjusting for known risk factors like blood pressure, blood glucose levels and fats in the blood," Dr. Lin said in a statement. "This could mean that perhaps there is some other protective mechanism that may be protecting those who eat a vegetarian diet from stroke."

Another Tzu Chi study showed reduced medical expenses among vegetarians compared with those consuming a diet containing meat. Researchers from Taiwan looked at more than 12,000 Buddhist volunteers and compared a diet with frequent consumption of fruits, vegetables, soy, and nuts to a diet characterized by relatively more consumption of meat and fish and less plant-based foods. Vegetarians had a 15% lower total medical expenditure and a 13% lower outpatient medical expenditure, compared with omnivores in this cohort. Vegetarians had lower expenses related to hypertension, dyslipidemia, depression, heart disease, and renal disease. The vegetarians also had medical expenditures that were 25% lower than those of the general population.

Chang Ya-lin, nutritionist at Tzu Chi Hospital in Taipei, cites a survey by the Journal of the American Medical Association which suggests that meat consumption leads to a loss of calcium and higher mortality rates. A four-year study of those eating relatively less meat but who switched to high meat diets, doubled the risk of getting

A selection of vegetables, fruit, mushrooms and grain makes for a tasty and healthful meal. Tzu Chi actively promotes vegetarianism and its members agree to follow a vegetarian diet. (Photographer / Huang Shu-Fen)

diabetes. The study's definition of low meat intake was two portions per week and high was seven or more portions. The survey also notes there is less of a strain on the kidney from eating less red meat.

Ms. Chang says the hospital (which has about 1,000 beds and was 90% full at the time she was interviewed in late 2019) serves only vegetarian meals. "This is for the health of our patients," she says. It serves three meals a day though only about 51–55% of the patients order meals during their hospital stay. Patients are free to eat their own food brought in from outside, but there are reminders at the hospital entrance and in a guide for new patients that the hospital is vegetarian only.

She says there is no need to worry about insufficient iron and calcium from not eating meat, as this can be offset by eating more beans and nuts. "If you switch to soybeans and nuts, you will see the effect. Also, use citrus fruit to increase the rate of absorption of nonheme iron from vegetables, nuts and beans."

The hospital tries to follow Tzu Chi's guidelines of using seasonal vegetables and buying those that are locally produced to ensure better health for staff and patients and reduce the need for long-distance transport. Prices are kept low in the cafeteria to encourage staff to take their meals there (NT$35 — or about US$1.10 — for lunch).

There are other reasons for maintaining a vegetarian diet. Plant-based diets have a significantly lower impact on the environment. According to the Food and Agriculture Organization (FAO) of the United Nations, 14.5% of anthropogenic greenhouse gas emissions, including methane, are from global livestock. Cattle account for 65% of livestock emissions. In addition to producing methane, cows consume large amounts of water; it takes an estimated 15,515 liters of water to produce 1 kg of meat.

Rising affluence causes a shift to diets that include more meat and dairy products. By 2050 the world's population is expected to rise 15% to over 9 billion people and global demand for meat is likely to climb 73% by then, according to the FAO.

According to some estimates, more than 70 billion land animals are raised and slaughtered each year to provide food for humans (poultry accounts for the bulk of that total). About one-third of global arable land surface is used for livestock feed production. Dietary changes could free up several million square km of land and reduce global CO_2 emissions by up to 8 billion tons a year compared with business as usual practices.

It is estimated that transitioning to more plant-based diets, in line with recommendations on healthy eating by the World Health Organization, could also reduce global mortality by 6–10% and food-related greenhouse gas emissions by 29–70%. A transformation to healthy diets by 2050 will require substantial dietary shifts, including a greater than 50% reduction in global consumption of foods such as red meat and sugar and a more than 100% increase in consumption of healthy foods such as nuts, fruits, vegetables, and legumes. This could prevent up to 11.1 million deaths per year in 2030, a 19.9% reduction of all premature mortality due to prevention of cardiovascular disease, diabetes, and cancer, among other diseases.

"As Master Cheng Yen says, we want to co-exist with the Earth. In doing that we accept that all living beings are equal. But when we (at the hospital) talk of vegetarianism to our patients, we emphasize the health reasons," says Ms. Chang.

There are differences in the response from different segments of the general public, however. "Generally, younger people are more receptive. They care about the environment and protecting animals. In many cases, they drink alcohol, but they still choose vegetarian food. For many older people, however, vegetarianism is a religious conviction."

Asked whether vegetarianism ever creates problems for the medical staff from hospital patients, she concedes that sometimes it does. "I have had problems with patients who insist on eating meat," she says, adding that in some cases the resistance is very much misplaced. "One patient with gout felt a flare up was due to the vegetarian diet. In fact, it was arthritis."

Tzu Chi works to promote a vegetarian diet at all of its seven hospitals in Taiwan, as well as its medical operations elsewhere. At the Taipei hospital it is calculated that vegetarian meals served to patients and staff save 421 metric tons of CO_2 emissions a year. Vegetarian meals are also being promoted across Tzu Chi's operations; if all of Tzu Chi's global volunteers are counted, vegetarian meals would result in a savings of about 92,000 tons of CO_2 emissions a year.

In order to achieve an even greater impact, Tzu Chi launched Ethical Eating Day in 2014. Under this ambitious initiative, Tzu Chi enrolled people to pledge to adopt a vegetarian diet for one day each year. They chose January 11 in a worldwide event to promote awareness of climate change and ending cruelty to animals. In 2019, more than 1.3 million people around the world pledged to observe a strict vegetarian diet for the day.

Tzu Chi volunteers prepare a vegetarian meal. Plant-based diets have substantial health benefits and a significantly lower impact on the environment. (Photographer / Xu, Jin-Fu)

Chapter 7

Reducing the Carbon Footprint

Preserving Life, Health and Love

While the mission of the Tzu Chi hospital staff is to prevent and treat disease, the charity defines health in a broad sense. The medical care mission also includes protection of the environment, minimizing the use of the Earth's resources and recycling.

"Our basic principle is not only treating disease. We also promote health," says Dr. Ming-Nan Lin, vice superintendent and director of the Department of Family Medicine at Tzu Chi's Dalin Hospital. "For health you need to think of risk factors such as the environmental impact from things such as toxic hospital waste. We need to respect not only people but also Mother Earth."

In 2019, Tzu Chi completed its seventh Taiwan hospital in Douliu in the south-central part of the island. (It also has a stem cell center in Hualien and its Suzhou Tzuchi Clinic, China) Data show that for the six hospitals in operation in Taiwan in 2018, the Tzu Chi hospital system served 3.78 million outpatients, with 240,000 emergency treatments and a total of 130,000 hospitalizations that year.

Tzu Chi spent NT\$10.36 million (about US\$370,000) between 2016 and 2018 on reducing the carbon footprint of its hospitals as it expanded its effort to meet the need for medical treatment in Taiwan. It estimates that these measures helped save 10,692 metric tons of CO_2 equivalent over that period.

Tzu Chi's hospital in Taichung makes use of solar power to lower its carbon footprint. (Photographer / Lin Yan-Huang)

In reducing electric power consumption, Tzu Chi hospitals used some measures as simple as painting the rooftops white to reflect light and reduce heat. They also switched to LED and T5 long life lighting, put timers on outdoor lighting, installed power saving equipment to control air conditioning, and set computer screens on "sleep" mode.

They water plants and flush toilets with reused water, which accounts for 17% of the total water consumed. At Tzu Chi's hospital in Taichung, the results are even better as rainwater and reused tap water together account for about 30% of total water consumption.

But the biggest contributor to reducing the carbon footprint of this big medical operation is the promotion of a plant-based diet throughout the entire hospital system. About four-fifths of the reduction of greenhouse gas emissions is a result of the reliance on a vegetarian diet for staff, patients and visitors. Only plant-based food is served in the Tzu Chi hospitals and that includes the staff cafeteria as well as the outside vendors brought in to provide meals for outpatients and

Volunteers put ecologically friendly paving stones into place at Tzu Chi's Taipei hospital. The paving stones allow better drainage and greater air circulation, reducing the environmental impact. (Photographer / You Cai-Xia)

visitors. Vendors must accept Tzu Chi's requirement to offer only plant-based food. The hospitals, which have a combined staff of 9,594, served a total of 3.6 million vegetarian meals in 2018, or a savings of 2,811 metric tons of CO_2 equivalent.

There is a recycling operation at the Dalin Tzu Chi hospital campus and all new staff undergo an orientation program that includes recycling. Throwaway chopsticks are shunned, and plastic straws have been replaced with paper straws. The use of paper cups is discouraged by offering discounts to those using porcelain cups. The hospital system recycles plastic IV bottles and plastic bags. While some waste such as toxic biohazard waste (blood-contaminated protective equipment) must be incinerated according to legal requirements in Taiwan, there is room for the recycling of many other materials, including confidential documents, for example, which can be shredded and turned into pulp with the help of an industrial partner. As a result, the hospital system had raised its recycling rate to 37.1% from 33.5% as of 2016.

Green Space, Green Materials

"Cherish the Earth" is much more than a slogan throughout the Tzu Chi organization. That holds true at the Tzu Chi University of Science and Technology, which seeks to instill in its more than 3,000 students, faculty and staff a sense of the importance of environmental protection. The attention given to the environment is hard to miss at the university, which has 155,515 square meters of green space, or about 88% of the total campus grounds. There are more than 3,000 trees, including Bodhi trees, red cedar, cinnamon, Formosan ash, magnolia, and banyan bread trees, while there is a wide variety of flowering plants such as orange jasmine, hyacinth, and mock limes among others.

For first-time visitors, the initial impression is the campus is exceptionally clean despite the fact that there are few garbage bins and only a handful of full-time cleaning staff. Students are expected to sweep up their own dormitories and keep the campus tidy. In order to throw away a simple piece of paper — such as a candy wrapper — a hike to the recycling center is required. This is a constant reminder to students and faculty alike not to create waste.

All incoming students are required to attend lectures on the environment and participate in recycling operations. The university also strives to encourage energy-saving behavior, such as turning off lights and shutting down computers when leaving a room and keeping the use of air conditioning to a minimum. A logbook keeps track of whether lights and computers have been turned off when a classroom is no longer in use. The logbook entry must be signed, signifying that someone takes responsibility for its accuracy.

"Ever since we set up the university, we have called on students to clean up their campus including their classrooms," says Niu Chiang-shan, dean of academic affairs. "If students enter a clean campus and they throw away a single piece of paper they will feel they are bringing down the standard of cleanliness. The environment around a student is very important. Students may not notice it, but they absorb this behavior and it becomes part of their life."

Over the years the administration has employed a number of other measures to reduce the university's carbon footprint. Green materials were used in construction whenever possible. In 2018, the university spent NT\$3.58 million on new green equipment and materials. New green equipment includes energy-saving T5 or LED light bulbs while other outdoor lighting is solar-powered and buildings are connected by a walkway covered with extra thin solar panels. This reduced electricity use to 1,358.54 kilowatt hours per person from 1,486.47 kwh per person the previous year. The university recorded a big reduction in water use to 169.15 liters per person that year from 204.95 liters the previous year. It made use of rainwater and turned kitchen waste into compost for the gardens on the university grounds. The university also recycles 54% of its waste.

"There is a long history at the university of promoting a simple life," says Chen Han-lin, who teaches environmental studies. "There are lots of activities outside the classroom to reinforce our connection to the environment. Students need to join at least five of them per semester. The idea is to make this part of their life."

In addition to offering 10 courses in environmental issues, the university organizes competitions with environmental themes in order to reinforce the environmental messages. Nearly 1,400 people took part in campaigns to clean up campus streets, paths and ponds over the year.

On campus, food at the cafeteria is vegetarian and nearly 193,000 meals were served to teachers and students over the course of 2018, reducing 153 metric tons of CO_2 equivalent. Prices are kept low to encourage students to eat on campus and maintain a healthy diet. The university also tries to ensure that local fruits and vegetables are used wherever possible by reaching out to local farmers.

The university produces 1.7 kg of waste per person per month or 0.06 kg per person per day, well below the national average, as measured by the Environmental Protection Administration, of 1.139 kg per day.

This sustained effort to encourage a more considerate use of global resources could also have benefits beyond Taiwan. "We have many foreign students and they are spreading the seeds (of a simple lifestyle) abroad," says Dean Niu.

Organic Farming and Poverty Reduction in Indonesia

The Tzu Chi University of Science and Technology has a specially designed a program with Indonesia to spread the message of a simple

Students from Indonesia at Tzu Chi University of Science and Technology show off their organically grown produce. They are studying organic farming techniques under a joint program with Indonesian conglomerate Sinar Mas and the Al Ashriyyah Nurul Iman Islamic Boarding School in western Java. (Photographer / Cai Yin-Ying)

Indonesian students tend an organic garden at Tzu Chi University of Science and Technology as part of a program to encourage organic farming and contribute to poverty reduction in Indonesia. (Organic farming can keep more carbon stored in the soil. Production, transportation and use of mineral fertilizers contribute directly and indirectly to emissions of greenhouse gases). (Photographer / Cai Yin-Ying)

lifestyle. The university brings Indonesian students to the campus for studies covering organic farming and cultivation methods as well as marketing and distribution. It has teamed up with the Sinar Mas Group, a leading Indonesian conglomerate that has major operations in food and agribusiness, and the Al Ashriyyah Nurul Iman Islamic Boarding School in western Java, a school that has ties to the Tzu Chi organization that date back nearly two decades. Sinar Mas funds most of the 30 students in the three-year program. After completion of the courses the students go on to employment at the company.

"The classes cover both the theory and practice of organic farming," says Su Mei-hui, one of the teachers in the program. Other teachers from the Agricultural School at Hualien supplement the instruction. The participating students were selected from more than 500 applicants from across Indonesia and required to have either Chinese or English language skills. It was designed as a four-year program but condensed into three years with instruction extended through normal semester breaks.

The course of study was born out of an idea that improving farming techniques in Indonesia would promote poverty reduction in that country, according to You Kun-tse, the head of the program. Sinar Mas took an interest because some Tzu Chi members in Indonesia are executives of the company. Bringing better farming practices was viewed as a path to poverty reduction.

Although there is debate about the net effects of organic farming, in the long run it is seen as saving energy and protecting the environment by sequestering carbon in the soil, leaving fewer residues in food and avoiding pollution of ground water. Drawbacks include lower crop yields and added costs.

Students seem to have taken these ideas on board. "There are many advantages to using organic farming techniques," says Timothy Varian Wynn, a 19-year-old student from Surabaya who is part of the program. "While the (cultivating) costs are higher it is more sustainable and (results in) better quality produce. When I go back to Indonesia, I hope to introduce this technology. This will raise the level of technological skills, promote progress and provide healthful food."

Chapter 8

Education and Recycling

"Our recycling volunteers have become professors of recycling"
— Master Cheng Yen

About 60 students from the Min Yang kindergarten have come to the Tzu Chi Bade Environmental Education Station in Taipei for a little taste of what recycling is all about. They are seated around a group of recycling volunteers and kindergarten teacher Chou Chia-wei, who also happens to be a Tzu Chi volunteer. "Where does all this garbage come from?" asks one of the recyclers. "People!" shouts the class in unison. "And why do we recycle?" asks the volunteer. "To love the Earth!" is the reply.

These are core messages that Master Cheng Yen includes in her regular lectures broadcast on Tzu Chi's television programs, published on Tzu Chi's website and taken to heart by recycling volunteers. Convinced that environmental awareness needs to start at a young age, Tzu Chi has sought to extend its recycling message by reaching out to local schools and inviting them to participate in its educational programs, just like the group at the Bade Environmental Education Station.

The Min Yang kindergarteners are shown a wooden board that has recyclables hanging from it. They are labeled with an easy-to-remember phrase: "ping, ping, guan, guan, chi" or "bottles, bottles, cans,

Students at Tzu Chi University of Science and Technology take part in recycling activities. All incoming students are required to attend lectures on the environment and participate in recycling operations. Students are encouraged to take part in campaigns to clean up campus streets, paths and ponds. (Photo provided by Tzu Chi University of Science and Technology)

cans, and paper" — some of the most commonly recycled items. All are easily distinguished by the youngsters.

Another board has a slightly more complicated message with the word "dian" or electricity — mainly referring to used batteries — followed by the numbers 1–3–5–7 written across it. When pronounced, these sound like other types of recyclables. "Yi" — or one — sounds like the word for clothing, "san" or three stands for three types of electronic products or computers, home appliances and phones. "Wu" or five signifies metals and "chi" or seven sounds like the word for "others." An example of these items also hangs from the board, giving the youngsters a close up look at what kinds of castaways can be reused. "All of these can be recycled," says one of the volunteers, pointing to the discarded items.

The Min Yang kindergarteners nod in agreement, seeming to get the general idea. The message is reinforced with a short video, aided by cartoon characters and some whimsical sing-along music as the youngsters clap in time. Finally, they are asked to join in the recycling effort — a task they obviously have been waiting for — by sorting different types of materials. They jostle each other for the chance to pick up plastic water bottles from a pile on the floor and place them in a big bamboo basket.

"Easy. Easy. Don't throw things," says Ms. Chou, her voice rising with a faint trace of alarm. Ms. Chou turns to a visitor and says: "We are giving them practical experience."

The group then moves on to help sort different types of paper. The kindergarteners are encouraged to help tear pages out of discarded books and magazines which are destined for recycling. With gusto, they rip out pages and toss them into different bamboo baskets:

Students listen to a volunteer at a recycling station explain the importance of recycling. Recycling stations also serve as environmental education centers. (Photographer / Lai Xiu-Ju)

printed paper in one basket, cardboard in another, glossy magazine paper elsewhere. One of the kindergarteners, Liu Che-rui, who says he is "almost" five years old, is asked why he and his classmates are doing this. "Because we love the Earth," he says, proudly showing he has been paying attention.

"We are teaching them to embrace these ideas," says Hsiao Hsiu-chu, who is in charge of the recycling station. "We need to teach children that recycling is important. This needs to start with the young." Ms. Hsiao and her husband came to recycling by a circuitous route. They had run a company that distributed throwaway paper cups and plates as well as plastic bags. She says the business was quite profitable, but her views changed after she was introduced to Tzu Chi in 1999.

Students from the Min Yang kindergarten have come to the Tzu Chi Bade Environmental Education Station in Taipei for a little taste of what recycling is all about. Here they are shown listening to a Tzu Chi volunteer explain how plastic bottles are sorted and recycled. (Photographer / William Kazer)

"Master Cheng Yen was calling for us to protect the environment but what I was doing was just the opposite. I found it harder and harder to continue this way."

Eventually, she got out of the business and fully embraced the cause of protecting the environment. "This was practical, rewarding work. It taught me so much." Now she is doing her part in this green movement to reduce waste and spread the word about the need for protecting the environment.

"We have visits from children from primary school to high school and college," says Ms. Hsiao. Besides students, there are other groups as well. Despite continuing political friction between the governments in Taiwan and mainland China, groups from Guangzhou to Chongqing and remote Qinghai province visited the recycling operations at the Bade station over the course of 2019, according to the station's record books.

These groups were looking to see if they could duplicate Tzu Chi's success back home. "They come to see how we sort recyclables," she says. "They want to see the Tzu Chi operation."

As many as 1,500 visitors came to see the Bade recycling operation in 2019, though visits in 2020 were initially affected by the Covid-19 pandemic.

In addition to class instruction, the Bade recycling operation has another unusual tool for teaching about preserving the environment. It is home to a "recycling museum" — a collection of household items that once had an important place in someone's daily life but eventually were considered obsolete and discarded. Old clocks, suitcases, sewing machines, rotary telephones and an Underwood typewriter — all have a place in the collection alongside ceremonial swords and coins from imperial China. Each item has its own QR code. "Many people would throw these things away, preferring something new," says Ms. Hsiao. "But I saved them. I think too many youngsters don't know what life was like in the past. In the past people had much less and repaired what they had. There was far less waste."

The Bade station is not alone in the Tzu Chi organization in promoting environmental education. Many of Tzu Chi's other recycling centers are also actively engaging with the community in the effort to raise awareness of the need to conserve resources. At the large recycling center in the Sanchong district in nearby New Taipei City they too have a steady stream of visitors. "We have about 150 students scheduled to come here in the next month," says volunteer Luo Mei-chu. She adds that students need to learn the importance of making do with less. "There is so much waste. If we don't recycle we will sink under the weight of all of this rubbish."

In December of 2019, the Kaohsiung Tzu Chi Jing Si Hall received certification from Taiwan's Executive Yuan as an Environmental Education Infrastructure Site, a highly coveted designation. An exhibition hall at the site, which is open to the public, has displays on global warming, a "low carbon footprint lifestyle" and "merciful technology" that helps protect the environment. It also provides courses on a variety of environmental subjects and these qualify as required study for students as well as government officials. Under the amended Environmental Education Act of 2017, officials, corporate executives and students are required to attend at least four hours of classes on environmental issues a year.

Chen Che-lin, who heads up Tzu Chi's environmental educational efforts in Taiwan and spent three years in mainland China building up educational operations there, says operations in Sichuan province received certification as qualifying for environmental instruction from the provincial and national authorities. Between 2011 and 2018, more than 80,000 people — from schoolchildren to government officials and members of non-governmental organizations (NGOs) — toured its educational facilities or attended lectures on the need for environmental protection and recycling (see Chapter 10). The Environmental Education Act also requires these organizations to designate people who will map out a program for environmental education.

Tzu Chi's environmental education efforts complement curriculum in Taiwan's schools. There has been school instruction in waste management and environmental protection since the 1980s. Waste management and recycling education in Taiwan ranges from schoolyard cleaning activities to waste material classification and practical experience in sorting to courses in environmental management and other related fields.

In 2001, the importance of environmental education was strengthened by a reform that integrated activities into a formalized curriculum for grades 1 through 9. This was a huge victory for environmental education scholars and practitioners, who had long sought a more structured approach.

In 2011, Taiwan's initial Environmental Education Act went into effect, enshrining earlier initiatives in an effort to cultivate public understanding of the environment and sustainable education. The Environmental Protection Administration, which was set up in 1987 to combine scattered regulatory responsibilities in a single government body, and the Education Ministry started to put together textbooks and devise teaching content for classes that started with kindergarten pupils. Environmental education personnel can be certified through education, experience, expertise, recommendation, examination, and training. Environmental educators in Taiwan can be trained in certified facilities or institutes.

However, in comparison with other studies, environmental education is a relatively new subject and further efforts are being made. The Ministry of Education, the Environmental Protection Administration and several NGOs have helped researchers and practitioners make environmental education a key priority. For example, the Chinese Society for Environmental Education in Taiwan has participated in the research, development, and promotion of environmental education since it was formed in 1993.

The amended version of the Environmental Education Act mandated that central government departments, larger corporations and schools all needed to designate staff who were responsible for drafting details of programs on environmental education. The required four hours of education courses for students, teachers and staff a year remained in effect.

"Passage of the Environmental Education Act was a very important step," says Hsu-Ming Yen, counselor and executive director of the Recycling Fund Management Board, which has control over a government fund used to promote recycling. "But this is something that has taken time."

Chapter 9

Community Engagement

Spurring Environmental Awareness

Every year, in the third month of the Chinese lunar calendar, Matsu — the patron goddess of seafarers — sets off on a pilgrimage from her home at the Jenn Lann Temple in Taichung's Dajia district. Tens of thousands of people accompany a statue of Matsu, which is carried aloft on a journey of nine days and eight nights covering some 300 kilometers. The Matsu tradition originated in Fujian in mainland China, and the pilgrimage traces the path of early immigrants in central Taiwan. On the way, Matsu stops at numerous temples in what is said to be a show of respect for the other shrines and deities important to the Taiwanese.

Firecrackers, traditional lion dances, the sounding of gongs, the blaring of horns and the burning of paper money are all part of the Matsu festival. It is one of the most important religious events in Taiwan and the associated customs and beliefs have been included by UNESCO as a "significant intangible cultural heritage of humanity."

The Matsu festival weaves the community together, and as part of that community, Tzu Chi is eager to participate. In 2019, Tzu Chi volunteers set up some 20 trash and recycling collection points which netted six metric tons of trash and recyclables. Tzu Chi volunteers were out in the early hours after the first day's procession, sweeping the streets clean of the litter and debris from firecrackers. Hong Li-tang, a

Tzu Chi volunteer for the past 30 years, says he has been participating in the Matsu procession for almost half of that time. About 300–400 Tzu Chi volunteers took part in 2019. "We had banners saying, 'Don't Litter, Recycle' and 'No Trash on the Ground'," he says, in a reference to a successful policy used in Taipei calling on residents not to leave trash bags sitting on street corners waiting for pickup.

Part of the Matsu tradition is to offer food and water to the faithful. In addition to carting away much of the trash, Tzu Chi demonstrates its commitment to reducing waste by supplying food stalls with cups, bowls and chopsticks, which it washes and takes home. "We had a message of 'Don't use throwaway tableware.' We work with many other groups and we will continue to do this. There is lots of work to do," says Mr. Hong.

Another area where Tzu Chi has made an important contribution to environmental awareness is the clean-up affecting another religious figure — Kuanyin, the Goddess of Mercy. More precisely, it was Kuanyin Shan, or Mt. Kuanyin, so-named because of its resemblance to the features of a reposing Goddess of Mercy. In 2017, many local residents concluded that mercy was in short supply there. Too many tourists had dumped trash along the mountain's pathways, creating an eyesore where a verdant wooded area had once been.

A middle school teacher who was dismayed over the unsightly state of Kuanyin Shan sounded the alarm and this caught the attention of workers at Da Ai TV. "A group of eight of us went to take a look," says Luo Heng-yuan, who was one of the coordinators of the clean-up campaign. "It took a lot of planning. We worked with the local authorities managing the scenic area."

One obstacle was that trucks couldn't get up to the peak to haul trash away. There was a steep climb of some 1,200 steps to reach the peak, which was 1,600 meters above the mouth of the Tamsui River below.

The first clean-up effort was launched on Earth Day in 2017. Tzu Chi volunteers were joined by many other groups — hikers, office

Volunteers dislodge trash from the undergrowth on the slopes of Mt. Kuanyin in a campaign to clean up the popular tourist spot near Taipei. The clean-up drive, begun on Earth Day in 2017, triggered a huge public response. More than 3,000 people took part in the effort. (Photographer / Huang Pei-Xiu)

workers, school kids and senior citizens. "There was a lot of publicity about the effort," says Mr. Luo. "It was a strenuous climb to get to the peak. So, we had three medical stations along the way just in case anyone needed assistance." Tea and refreshments were served as well. Some of the volunteers used mountain climbing gear to retrieve trash on one of the slopes. They hooked themselves to a rope-line for balance as they dislodged rubbish from the undergrowth along a steep slope that had long been used as a garbage dump. They gathered up plastic bottles and bags, cans, cookware, and bits of scrap metal among other discarded items. Volunteers formed a human chain to bring down the bags of garbage that had been recovered from the slope and elsewhere on the mountain. Others carried bags down the steps. In some cases, volunteers carried the trash bags slung over a pole across

Tzu Chi volunteers haul bags of recyclable materials collected on Pulau Ketam (Crab Island) in Malaysia as part of a clean-up drive. Recyclable materials were taken by boat from the popular tourist destination to nearby Port Klang for sale to traders. (Photographer / Jiang Run-Fu)

their shoulders like an old-style porter. Finally, the garbage bags were deposited in a parking lot below where they were loaded on trucks and carted off for disposal.

"People felt this was a really good cause," says Mr. Luo. "Lots of people responded to the effort."

While the operation was a success it was clear this was only scratching the surface. "We realized we were just getting started. There was much more trash to remove," says Mr. Luo. Ultimately it took three more days of hard work to carry down all the trash and restore the Goddess of Mercy to her former pristine state.

The group hauled out 7,250 rice sacks filled with garbage and weighing a total of 25.8 metric tons. It took 72 truckloads to cart the trash away. More than 3,000 people took part in the effort — including people who happened to run into the cleanup effort by coincidence and decided to join in.

"This effort spurred environmental awareness in Taiwan," says Mr. Luo. Other similar activities followed. In the Taoyuan area there was a community effort to spruce up Miaotou Shan in a drive that was inspired by the Kuanyin Shan campaign.

Tzu Chi has continued with other efforts to encourage the public to become more active in the campaign to protect the environment. In Yilan along the northeast coast, Tzu Chi has taken aim at cleaning up the waters around Taiwan. It made a proposal to fishermen at the Yilan port asking them to avoid dumping their trash at sea. Tzu Chi offered to collect the trash on their return to port. Volunteers gathered empty cans of food, water bottles and batteries. In six months, they had collected 200 kgs of batteries alone.

It is not only in Taiwan that Tzu Chi works with local communities in an effort to tidy up the environment. In Malaysia, Tzu Chi volunteers have been instrumental in kickstarting a cleanup of a popular tourist destination, Pulau Ketam or Crab Island, known for its fishing village and mangroves. Popularity with tourists has helped the island's economy but it has also contributed to problems with too much trash, much of it ending up clogging the harbor and polluting nearby waters. Tzu Chi volunteer Tan Sow Lang was the inspiration for cleaning up the Pulau Ketam environment. She would ride her bicycle around the island in her spare time, picking up garbage and recyclable material. Heng Bok Meng, who runs a fuel trading business, helped out as the local recycling effort took shape. He delivered the recycling material to Port Klang, the main port on Peninsular Malaysia, where it was sold to traders. The funds generated by the sales were used to support the needy.

Heng Bok Meng joined Tzu Chi as a volunteer and was sent to a training course in Taiwan, which was described as an eye-opener. "I became very involved after returning from Taiwan," he said, according to Tzu Chi's official website. "When I saw the elderly volunteers working very hard in sorting recyclables, I had no excuse for not doing it since I am still a young man." This demonstrates how individuals can make an impact and change community behavior.

Tzu Chi Foundation Malaysia adopted what it called the Environmental Protection Mission in 1995, starting in the state of Malacca and quickly expanding the campaign.

Tzu Chi volunteers had been promoting the concept of environmental protection by curbing one's desires and reducing consumption long before the Malaysian government launched its initiatives to raise public awareness of recycling. Tzu Chi volunteers had been going door-to-door to encourage the public to reduce waste and avoid discarding items that could be reused or recycled. They also set up small-scale community recycling points to encourage local residents to deposit their reusable and recyclable items there.

Tzu Chi has also joined the efforts to work with local communities elsewhere in efforts to encourage recycling and protect the environment. In Los Angeles, Tzu Chi volunteers have run programs to aid the homeless and low-income schoolchildren, and that has made it possible for Tzu Chi to introduce its recycling ideas. The method is similar to the approach used in other locations; Tzu Chi offers blankets — which use plastic bottles as a raw material — in return for plastic bottles collected. "Find 70 bottles and we give you a blanket," says Debra Boudreaux, a commissioner and Tzu Chi Foundation USA executive vice president and senior representative of the United Nations New York/Geneva office. One of the targeted areas was in Antelope Valley in California, where there is a large concentration of low-income households that are in need of assistance. "Some schoolchildren were living in cars," she says. As in other recycling efforts, the blankets were part of a program designed to encourage participation in environmental protection efforts. "We give you a blanket, but you have to earn it."

Tzu Chi has also worked with the faculty members of Western University to offer blankets to the homeless in exchange for collecting bottles. "The message is to try to change behavior," she says.

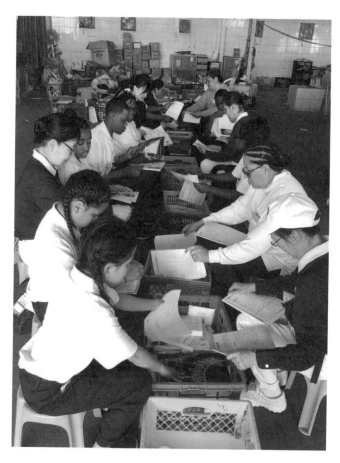

Public housing residents from San Francisco's Hunters Point neighborhood take part in sorting recyclable material during a visit to Taiwan. Tzu Chi has a number of programs designed to help local residents in need. The trip was organized by Tzu Chi in 2018 to give residents practical experience in recycling and train future community leaders in environmental protection. (Photo provided by Tzu Chi Foundation)

On March 1, 2020 New York State implemented a ban on single-use plastic bags. In an effort to help New Yorkers adjust to the new law, the day before the ban went into effect, Tzu Chi volunteers distributed 1,000 free eco-friendly reusable bags in the heavily Asian neighborhood of Flushing in New York City's borough of Queens. In addition to presenting a helpful reminder in advance, the distribution was used to bolster awareness of the impact humans have on Mother Earth — a key component in Tzu Chi's humanitarian activities internationally. New York State uses 23 billion plastic bags a year and New York City alone uses 10-billion of them. The ban was in response to public pressure amid mounting levels of plastic waste that left plastic bags clogging drains, stuck in trees or floating in waterways. Tzu chi participated in meetings with city officials on the implementation of the ban and frequently visited supermarkets in the city to advise them of the challenge ahead. Cashiers at supermarkets were instructed to remind customers that bags would no longer be free.

Hunters Point Project: Gratitude, Respect and Love

Tzu Chi's environmental educational efforts take many forms. In San Francisco, for example, Tzu Chi volunteers have partnered with the Malcolm X Academy, a primary school in Bayview-Hunters Point, one of the poorest areas of the city, to enhance existing instruction and nurture a recognition of the need for protecting the environment. In 2009, Tzu Chi was invited by the academy's principal to assist in classroom instruction and provide one-on-one after-school tutoring, particularly in math. Students learn about the environment and vegetarianism and study Mandarin Chinese as well. But behavioral and psychological issues have to be addressed in tandem with the classroom instruction. Some of the children come from broken homes while others are victims of parental abuse. There is a critical need for what Tzu Chi calls the ABCs — the basic requirement of Attendance, the need for proper Behavior and the goal of attending College. The Tzu Chi curriculum is also described as teaching "gratitude, respect and love."

After years of working with the Hunters Point community, Tzu Chi has won the confidence of the city government as well as local residents. It was tapped to run a food bank in the neighborhood,

Hunters Point residents listen to an explanation of gardening techniques during a visit to the Jing Si Abode in Hualien. (Photo provided by Tzu Chi Foundation)

distributing groceries to residents once a week. Students from the University of San Francisco pitch in, earning credits for their studies.

Tzu Chi also received a grant from San Francisco's environmental department and created a point system whereby residents of the Alice Griffiths public housing development in Hunters Point could take part in a recycling program. Students and adults earn points collecting cardboard, plastic bottles, and aluminum cans, among other items, to earn points that allow them to buy furniture. They put in 17 to 20 hours of work a month and get a Visa card that can be used for purchases. Children who qualify get US$150 while adults get US$200 as a monthly reward for community service.

Eligible participants must be residents of the Alice Griffiths public housing development in Hunters Point. A home evaluation revealed that some of the children in the neighborhood had no bed to sleep on, so Tzu Chi focused on getting beds for those families. The school kids also needed a quiet space so they could concentrate on their studies. Purchases through the point system also included a desk and chair with a lamp. Adults earned additional points by attending a workshop on recycling and the environment.

The program hasn't been all smooth sailing, however. "It's a high crime area and many Tzu Chi volunteers said, 'We don't want to risk our lives'," according to Roxanne Buchwitz, a Tzu Chi commissioner who has been a key member of the project. "But we asked Master Cheng Yen, who said Tzu Chi has a deep affinity with Hunters Point and the blue-and-white Tzu Chi uniform is bullet-proof'." (In the past Tzu Chi had considered building a Jing Si Meeting Hall in the area that was once home to a U.S. Navy yard. Redevelopment plans for the area have long been stalled due to complications related to a proposed toxic waste site clean-up).

In 2018, Tzu Chi organized a 10-day trip to Taiwan for a select group of residents to gain a better understanding of the charity organization and its accomplishments, learn more about recycling and environmental protection, and gain a broader perspective on life. Volunteers spent nine months preparing the selected students and community members for the trip. In some cases, they had to scramble at the last minute to locate a parent for permission to get a passport. "They had never been out of Hunters Point. It was a bit of a shock."

The trip was both an eye-opener and an inspiration for the participants who now are bringing the Tzu Chi spirit to their neighborhood. They also got to see for themselves that the achievements of DA. AI Technology, which recycles plastic bottles to produce blankets and other products, were real and not just part of classroom study.

Tzu Chi also has a college scholarship program which is now in its second year with the first recipient getting $1,000 per year. "We give them hope," says Ms. Buchwitz, speaking of the neighborhood schoolchildren. "We are able to give them a positive framework for their life."

Chapter 10

Recycling and Disaster Relief

"The wellbeing of the world should be everyone's responsibility"
— Master Cheng Yen

Tzu Chi has been on the front lines of relief work since its earliest days. An earthquake in China, war refugees in Turkey and Jordan, a typhoon in the Philippines, fires in California, and flooding in Indonesia. These are just some of the disasters — natural and man-made — where Tzu Chi has been among the first responders. It has tapped its army of thousands of volunteers to hand out food and cash to struggling disaster victims. It has built temporary and permanent homes, replaced collapsed school buildings, installed water purification systems and created employment in devastated neighborhoods.

But alongside its better-known disaster relief efforts, Tzu Chi has also been able to wield its considerable influence to promote another objective — addressing the need for protecting the environment. In many of the areas where Tzu Chi aids people in need, the assistance is required over a long period of time. That gives Tzu the ability to win the trust of these disaster victims and build a lasting connection. It also helps in the effort to convince communities to consider alternative patterns of behavior, particularly those that affect the environment.

Tzu Chi's efforts at rebuilding the lives of poor villagers on the outskirts of Jakarta are a case in point. In January of 2002 persistent

Children play on a swing at the Da Ai (Great Love) Village built with Tzu Chi aid after flooding along the Angke River in 2002. The model village, built on land supplied by the local government, has 1,100 housing units. Tzu Chi's continued links with the community have enabled it to promote recycling in the area. (Photographer / Yan Lin-Zhao 顏霖沼)

downpours caused severe flooding — the worst in three decades — which swept away more than a dozen people and forced widespread evacuations. In Kapuk Murua, a village along the Angke River, flooding was particularly serious. Household garbage thrown into the river by residents had clogged the waterway. Industrial waste, including heavy metals that were poured into the murky waters by nearby factories, also left the area with a record of chronic stomach and intestinal problems.

Alerted to the flooding, Tzu Chi Indonesia members toured the stricken area and saw residents camped out on the roofs of their inundated homes. They returned to the stricken neighborhood in boats to deliver cooked food and bottled water. They brought medical teams to take care of the sick. At times the boats were too big to navigate the clogged waterway, so they switched to smaller craft. Tzu Chi Indonesia asked for help from Master Cheng Yen, and Tzu Chi came up with a five-pronged approach that covered drainage of the existing floodwaters, cleaning up the stricken areas, disinfecting, providing medical assistance and undertaking reconstruction. They passed out school bags and stationery supplies to students but after a month they realized they needed a more active response. Some 350 medical volunteers from Tzu Chi Indonesia, Malaysia, Singapore, the Philippines, and Taiwan took part in the effort. Ultimately, they took up a collection to build new homes in another location for displaced residents. They made use of plans from the reconstruction effort in Taiwan following a devastating earthquake which killed more than 2,400 people and destroyed or damaged over 100,000 buildings in 1999. Indonesia provided five hectares of land with Tzu Chi taking on all construction costs. Tzu Chi built 1,100 apartment units in a model development called the Great Love Village. Besides apartments it has a school, shops, a Muslim prayer room and a recycling center.

The initial recovery efforts were also followed by a cash-for-work program to deal with an unemployment rate that was estimated at 50% initially. Tzu Chi provided vocational training and then provided employment opportunities.

"Now the surrounding area is really bustling so there are more jobs," says Hong Tjhin, CEO of DAAI TV Indonesia.

Tzu Chi also was successful in overcoming suspicions in the largely Muslim nation that its real intent was to spread Buddhism. Persistence, goodwill and the respect shown for other cultural and religious practices, helped overcome the early doubts.

After the villagers moved into the new and more comfortable apartments, there were continuing problems with garbage disposal. Recycling teams were organized, and the cleaned and sorted materials were sold to traders. Profits were used to support charity for low income households. "It has been a long learning curve but (the children) are the ones who are changing views on the environment," says the DAAI TV Indonesia executive. He adds that the schools built by Tzu Chi helped reinforce the message. Tzu Chi built a comprehensive

Tzu Chi volunteers organize a community recycling effort in Marikina, part of Metropolitan Manila. The city was hit hard by Typhoon Ketsana in 2009, and Tzu Chi organized a "work relief program" that brought reconstruction jobs to the stricken area. More recently, Tzu Chi signed an agreement with the city to promote recycling. (Photographer / Ma An-Chi)

primary, middle and high school at a new location farther away from the river. The high school alone, which takes in pupils from other communities as well, now has more than 2,000 students. These schools have been instrumental in creating a sense of environmental responsibility. Children at the village school are taught to separate garbage and recyclables. Starting from the lower grades, students become aware of the need for protecting the environment.

Rebuilding in the Philippines

Tzu Chi had a similar experience in the Philippines where it played a key role in providing relief and returning lives to normal after the

Tzu Chi relief volunteers arrive in one of the areas in the Philippines devastated by Typhoon Haiyan, one of the most powerful typhoons on record. Haiyan left 10,000 deaths in its wake, mostly in Tacloban and Ormoc in Leyte province in the Eastern Visayas in 2013. (Photographer / Zhong Wen-Ying)

devastation of Typhoon Haiyan in 2013. Typhoon Haiyan, one of the most powerful typhoons on record, brought widespread destruction to the Philippines, leaving some 10,000 deaths in its wake, mostly in Tacloban and Ormoc in Leyte province in the Eastern Visayas. Power was cut and some two million people were affected. Lucy Torres, an elected representative of Ormoc, described the city's structures as about 95% damaged or destroyed. Tzu Chi constructed 1,585 homes for some 6,000 people on land contributed by Ormoc's mayor. It handed out emergency funds and pumped cash into the regional economy by hiring local residents to cut bamboo for use in construction under a cash-for-work program.

Years later Tzu Chi has been able to build on the goodwill created in the aftermath of the typhoon. It organized recycling campaigns that also provided paying jobs to some of the low-income families in the area. Tzu Chi volunteers first did an assessment of living conditions and selected candidates for its cash-for-recycling program. They conducted door knocks to advise villagers of the recycling drive and its dual objectives — conserving resources and helping the less fortunate. They used the blankets made from recycled plastic bottles as a means of reminding people of how recycling allows Tzu Chi to assist those in need.

"I received (blankets) in the past as a care recipient so I understand that recycling is important and it really does help people," said villager Lerio Capuyan, according to Tzu Chi's official website. Another villager, Pilar dela Costa, says she was contributing her recyclables as a way of repaying Tzu Chi and her community for similar help in a period of adversity. "They helped improve our lives to where they are today," she said.

After the rebuilding was completed, Tzu Chi reinforced its message on the need to protect the Earth. It promoted dietary changes by means of its Ethical Eating Day — to encourage the following of a vegetarian diet as part of an effort called "new homes and new ideas."

More recently, Tzu Chi signed a recycling agreement with the city of Marikina in the Philippines to promote recycling, focusing on its schools. Marikina's mayor led a team to Hualien to present the agreement to Master Cheng Yen in person and thank the foundation for its help. With a population of 480,000, Marikina is one of the 16 cities that constitute Metropolitan Manila. In September of 2009, it was devastated by Typhoon Ketsana (also known locally as Tropical Storm Ondoy). The foundation decided to concentrate its relief effort in Marikina and employed over 5,000 residents in a "work relief program," under which participants were paid to clean up the mud and debris left by the typhoon. The foundation's work was seen as crucial to bringing life back to normal in the city and it inspired more than 1,000 local residents to undergo training to become certified Tzu Chi volunteers.

Sichuan Earthquake

In 2008, a massive earthquake in China's Sichuan province killed 70,000 and injured 375,000. It was so powerful that it shook

Tzu Chi relief workers distribute food after the massive Wenchuan earthquake, which killed 70,000 people in China's Sichuan province in 2008. Tzu Chi rushed aid to the stricken area, arranging deliveries of tons of relief supplies. Coupled with aid given after another earthquake in Ya'an in Sichuan in 2013, Tzu Chi distributed some NT$2 billion — or about US$68 million — in Sichuan disaster relief alone. (Photographer / Wu, Bao-Tong)

buildings in Beijing — more than 1,900 km away. Tzu Chi rushed aid to the disaster area, quickly arranging deliveries of tons of badly needed relief supplies. Damage to schools was extensive — nearly 7,000 of them were affected, many of them reduced to piles of rubble. Many of the killed or injured were schoolchildren and teachers (estimates run as high as 10,000). This aroused public anger and allegations of corruption that enabled shoddy construction of public buildings. Tzu Chi moved to restore permanent schools, constructing 13 in the disaster area. Over a five-year period it handed out aid and blankets as well as scholarships for more than 12,000 students. Tzu Chi also made winter distributions of blankets, heavier clothing, food and daily necessities to low income households, mainly in rural areas. This has continued years after the earthquakes rocked the region.

Coupled with aid given after another earthquake in Ya'an in Sichuan in 2013, Tzu Chi distributed some NT$2 billion — or about US$68 million — in Sichuan disaster relief alone.

Tzu Chi's efforts to rebuild those schools — as well as homes — has earned the heartfelt thanks of communities across the stricken area. It has also helped build an awareness that the environment needs protecting. In the Tzu Chi schools, students learn of the need for recycling. At the Luoshui village primary school in Shifang in Sichuan, the first of these 13 schools built by Tzu Chi (completed less than two years after the earthquake), instruction in the importance of protecting the environment starts with some of the youngest schoolchildren. Three times a week the second graders, accompanied by a teacher, carry blue plastic bins for recyclables such as plastic bottles and paper to the nearby recycling station. "I hope that the students remember at all times that we are all residents of this planet and need to cherish the Earth," said Li Zemin, a teacher at the primary school, according to the Tzu Chi website.

Luoshui's recycling center is now one of 38 Tzu Chi-built schools in mainland China. (According to Tzu Chi's published data, the charity has 6,000 volunteers, 44 permanent recycling stations and 323 temporary recycling sites in mainland China). Tzu Chi has been able to engage local communities and that has been particularly evident in Shifang and other parts of Sichuan. Wen Xurong, who runs a street stall serving fried pancakes, can attest to the beneficial effects of this assistance. Thanks to Tzu Chi's recycling activities, her husband Dai Anhua has found a more productive substitute for his regular mahjong pastime. In the past, he'd borrow money to play mahjong — and usually lose it. "It was really annoying. We'd argue and sometimes even fight," she was quoted as saying in an article on the organization's website.

But after he joined the Tzu Chi recycling effort his personality changed, according to Ms. Wen. Now her husband helps prepare for the day's business in the morning and rides a tricycle around the village collecting recyclables in the afternoon. He plays recorded music as he

Students from a primary school in Luoshui in China's Sichuan province carry a rubbish bin to the school's recycling center. Students learn at an early age how recyclables are separated from ordinary trash. The school is one of 13 built in Sichuan with aid from Tzu Chi after devastating earthquakes rocked the province. (Photographer / Bian Jing)

goes, alerting villagers that he is making the rounds in the neighborhood.

"They hear the music and know we are coming," he says.

Many of those who received help after the Wenchuan earthquake joined the ranks of Tzu Chi's recycling volunteers or helped with relief efforts. "Many of them are now our volunteers," says Huang Chong-fa who arrived in Sichuan in 2008 to join in the relief effort after the massive earthquake. These days he is still in Shifang. "They also brought their experience to the relief efforts in the Ya'an earthquake," he says. "They were able to encourage people to persevere. They demonstrated that since they had recovered from a severe earthquake it was possible for others to recover too."

This dovetails with Tzu Chi's goal of having those who receive help give assistance to others. It builds trust along with community cohesion and helps reinforce self-confidence among those who need to rely on aid from others.

In Sichuan Tzu Chi has also made use of its recycling efforts to find employment for local residents. As part of its recycling program, over 100 people are employed to take apart old refrigerators, TVs and other home appliances to recycle useful materials, according to Chen Che-lin, who spent three years working in China after the earthquake and now heads environmental education operations in Taiwan. The plastics and metals are sorted and sold to traders in the region. In Chengdu another smaller operation employs about 30 people processing materials used in making blankets. Business is conducted by DA.AI Gan En (Great Love and Grace) Eco-tech, a company set up by Tzu Chi. Profits from the recycling operations are used to pay staff and make donations to low-income families.

Fang-Tsuang Lu, director of the Charity Development Department of the Tzu Chi Foundation, said in a speech that children in the affected areas were enlisted to offer care to the displaced people. They first underwent training in giving massages, helping with cleaning and even providing light entertainment to ease the physical and psychological stress of the disaster victims. These children later told their parents of their experience. In many cases their parents joined the relief efforts, volunteering in the distribution of food. "We had one group taking over from the previous group and in so doing they brought the concept of protecting the environment to the disaster area," he says.

"They understood the need to be gentle with the environment and leave behind a clean and pure Earth for the next generation...They too promoted the idea of protecting the environment and thereby cast off the shadow of the disaster. As they gave of themselves, they found pleasure."

Integrated Operations

Tzu Chi has deepened the integration of its operations to leverage its existing strengths and carry out its core missions to better serve the community. The Tzu Chi liaison office in Miaoli is a prime example, combining recycling and long-term care operations with environmental and recycling education as well as disaster prevention training. The Miaoli office also includes displays on environmental protection and recycling as well as technology exhibits.

As environmental degradation and global warming contribute to the increased frequency and intensity of natural disasters, the combination of disaster prevention and environmental protection makes sense. For Tzu Chi, disaster prevention training includes the prevention of accidents that commonly affect the elderly. As Taiwan is already an aging society, the government has rolled out an elder care policy designed to enhance the quality of life of senior citizens. Tzu Chi, Taiwan's first non-government organization to play a role in disaster relief, worked to coordinate its activities with those of the government. While the government looks at prevention from a macro view of dangers, Tzu Chi takes a humanistic approach to individual households. Besides caring for the mental and physical health of the elderly, it addresses safety in the home. According to Louie Lu, head of Tzu Chi's disaster prevention unit, volunteers need to review safety in the home and determine whether there is adequate preparation. They assess whether the elderly have grab bars in the bathroom, whether floors have non-slip surfaces, and whether there are potential dangers from steps at the front door or inside the apartment. They also check to see if there are fire alarms and ensure that gas water heaters and electrical appliances are being used in a safe manner.

"In addition, when disaster strikes, we need to pay attention to the fact that these people are generally weaker and at greater risk. So, we need to ask whether there is a plan in place to take care of the elderly in the neighborhood," says Mr. Lu.

"From the government's perspective, the focus needs to be directed more towards preventing risks to the overall environment, but Tzu Chi is a civic organization with its roots in civil society and with a humanistic spirit," he says. "As we train people in disaster prevention, our fundamental principle is humanity first. We are primarily concerned with caring for the elderly and vulnerable, so we train up even more enthusiastic neighborhood volunteers... This is central to Tzu Chi's decision to join in disaster prevention training."

The Miaoli center completed its first disaster prevention training session earlier in 2020 and 20 Tzu Chi workers from around Taiwan participated. An application for certification of its training operations is being made to the disaster response association. Plans call for some 600 volunteers to be trained by the end of 2020.

At the Miaoli center, the foundation has pre-positioned a mobile kitchen which can serve 900 meals in three hours for rescuers as well as those rescued in emergency situations. It is mounted on a 3.5-ton open truck and includes a water purifier and cooker that can prepare meals without the need for bottled gas. When not in use for actual disaster relief, the equipment is part of the disaster relief training program.

Tzu Chi has also been cooperating with Jianxing University for training in the use of drones for aerial photography in disaster areas and it has signed an agreement with the National Science and Technology Center for Disaster Reduction (NCDR) to make use of its information system. Workers in the field can report back to headquarters, giving detailed assessments of disaster situations, making use of real-time information and incorporating data from the meteorological bureau, water department and the Interior Ministry. The system makes use of the NCDR's account on Line, the voice and data app that is widely used in Taiwan, Japan and South Korea.

Chapter 11

Toward Zero Waste — Technology and the "Bottle-to-Blanket" Program

An essential element in recycling is finding a new use for discarded materials. While Tzu Chi had made significant strides in recovering waste it needed to find new ways to use the recovered materials. The time-consuming process of sorting and cleaning resources makes sense only if there is sufficient demand for the recovered materials.

That eventually resulted in the creation of two companies that try to create eco-friendly products that can fill in some of the gaps in the circular economy.

Tzu Chi's DA.AI Technology Co., Ltd was the first of these green supply chain companies to take shape. DA.AI Technology was founded in 2008 with the aim of applying technology to turn used materials into eco-friendly products. All of the company's net profits are used to support Tzu Chi's charity work.

The founders were executives who wanted to put their business experience to work for the public good and for the cause of reducing waste and protecting the environment.

"We were looking for ways to sustain funding for our good works when the economy takes a downturn and people have less money for charitable donations," said James D.M. Lee, one of the company's founders and now the company's president/CEO.

Tzu Chi had already set up the Tzu Chi International Humanitarian Aid Association in 2003 with the aim of inspiring business leaders to contribute their knowledge and experience to the organization's disaster relief efforts. They looked at food, clothing, housing, transportation, information and communication products that could be used in international relief missions.

Mr. Lee says that the same business leaders had fresh motivation for collective action after a massive earthquake off Indonesia's coast triggered a devastating tsunami on Boxing Day in 2004. That killed

Disaster victims in the Philippines show their gratitude for timely assistance, expressing thanks to "God, Tzu Chi and the Red Cross." The blankets they are holding are products of Tzu Chi's "bottle-to-blanket" technology. Plastic bottles are shredded and used as raw material for woven fabric that is made into lightweight blankets. (Photographer / Bo Li-Ni)

some 200,000 people and inflicted widespread destruction around the Asian region and as far away as East Africa. Tzu Chi was instrumental in providing disaster relief, dispensing food and medical assistance and helping rebuild shattered communities. In the years since the company was inaugurated, it has become closely associated with Tzu Chi's humanitarian assistance.

The Tzu Chi International Humanitarian Aid Association was particularly concerned that non-biodegradable plastic bottles were overwhelming the environment. Reducing demand for these items and extending their product life were paramount objectives. In 2006, it began gathering vendors to set up assembly lines, using the recycled plastic bottles that Tzu Chi volunteers collected to produce eco-friendly textile products.

In the years since DA.AI Technology was established, the company has worked with a number of manufacturers that can use the waste materials collected by the Tzu Chi environmental protection volunteers to manufacture a wide range of products. The company's most commonly used products are the eco-friendly blankets made from recycled PET bottles. This is what Tzu Chi refers to as its "bottle-to-blanket" concept. Tzu Chi has distributed more than 1 million of these lightweight blankets to victims of natural disasters and political conflict as well as other people in need in some 30 countries — from Syrian refugees in Turkey and Jordan to victims of typhoon devastation in the Philippines and forest fires in California. It takes 76 plastic bottles to make one $70'' \times 90''$ blanket. The company and its partners shred these PET bottles into chips which can be spun into a fine yarn which can be woven into blankets and other products.

"These blankets also create respect for Taiwan and Tzu Chi," says Mr. Lee.

Technical breakthroughs allow a longer thread and ultimately a lightweight and softer blanket. DA.AI's innovative production process has received numerous awards — both local and international. Its production process has received a "Global Recycling Standard"

("GRS") certification from the Peterson Control Union of the Netherlands. In order to gain GRS process certification, all stages of production, from obtaining raw materials to manufacturing and trading of the products have to be in compliance with GRS standards. The eco-friendly fleece blanket made with 100% recycled polyester has also received Carbon Footprint Certification from TÜV Rheinland, a testing service and certification company based in Cologne, Germany, and recognition from Taiwan's Environmental Protection Administration.

DA.AI buys anywhere from 1,800 to 2,000 tons of PET bottles each year from Tzu Chi to produce blankets and other items. It has a range of other products such as shirts and trousers for Tzu Chi uniforms, polarized sunglasses, suitcases and other items.

"We are looking for products that are for daily use," says Mr. Lee, pointing to the company's uniforms and other clothing items. It also has also introduced footwear with tough plastic inserts that protect the feet of volunteers from puncture wounds as they operate amid the rubble of disaster zones as well as suitcases and hard plastic polarized sunglasses.

"Everything I am wearing has been recycled," says Mr. Lee. "We are slowly making progress and reducing the burden on the planet."

DA.AI started its own research and development division in 2012 to find new uses for recycled PET. It has teamed up with Taiwanese research groups like the Plastics Industry Development Center on new products, such as using bottle caps to make office supplies. And its researchers have worked on taking products already made from recycled materials and breaking them down to reuse as other recycled products, such as disaster relief blankets. Worn out blankets and textile fabric can be recycled in what the company calls its "Recycle To Recycle" effort.

"We try to invest at least 10% of our revenues to develop green technology or new eco-friendly products," said Lori Chen of the company's research and development operations, speaking at an international forum in Poland.

A Tzu Chi member at an outdoor display in New York explains what the charity organization does and how blankets used in disaster relief are made from recycled plastic bottles. (Photographer / Weng Xiu-Chun)

DA.AI has been certified as a B Corp., a measure of a company's corporate social responsibility. DA.AI says it made its effort to attain that status in order to benchmark the sustainability of its operations. The company donates its profits to the Tzu Chi Foundation or re-invests them in research and development.

In 2014, the company launched its anti-puncture footwear made from used PET bottles. These tough shoes were used by volunteers in disaster relief following earthquakes in Nepal in 2015 and 2019. They proved to be extremely useful for work amid rubble after earthquakes and devastating storms toppled buildings, exposing nails, glass shards and other sharp objects. In the relief efforts in the aftermath of Typhoon Haiyan in the Philippines, volunteers made use of the company's Eco Solar LED Cap. These came in particularly handy for relief work in darkness after the typhoon knocked out power in the region.

After a powerful earthquake struck Nepal in 2015, Tzu Chi sent in disaster relief teams, providing medical and other assistance. Chien Shou-hsin, head of the Taichung Tzu Chi Hospital, is shown here examining a Nepalese patient, using a cap with a solar-powered LED light. The product was designed by DA.AI Technology. (Photographer / Luo Rui-Xin)

"There are other companies that can make use of technology," says Mr. Lee. "What makes us different is our Great Love for humanity," he says, noting that company executives put in long hours without pay because of their sense of mission. "I don't believe any of us would do this for free at a for-profit company."

The company's chairman Huang Hua-te takes the same altruistic approach to the company's mission, noting it is unlike traditional for-profit companies in other ways. In the book *MIT — Made in Taiwan/ Tzu Chi*, Mr. Huang was quoted as saying: "The most important thing for us is not spurring sales of our products and certainly not encouraging more consumption. We play a small part in helping the environment by cherishing the Earth's resources and making use of recycling technology to give a new life to materials."

Huang noted that the company also makes minimal use of dyes for its textile fabrics. It combines transparent plastic bottles and carbon black, producing a gray blanket and using less water in the manufacturing process and cutting down on pollution.

"Many people can make products like these, but we expand the love that goes into them. In so doing we are reducing the burden of the Earth," he says.

Reducing consumption and waste is a key goal, according to Mr. Lee. "We start with 'don't use things or use less of them'," he says. "We want to get to zero waste and we call this a 'return on influence' rather than a 'return on investment'. At the same time, it elevates the value of the life of our recycling volunteers."

Still Thoughts

Jing Si (Still Thought) Pureland is another Tzu Chi company that aims to find more uses for recycled materials. It is technically under the religious association rather than the Tzu Chi Foundation and has operations that range from publications and bookstores to food and what it calls "eco-products."

Eco-products have played an important role in recycling and environmental protection as well as relief efforts. Working with Tzu Chi's recyclers, the company has been able to turn the omni-present paper lunch boxes, which have a thin layer of plastic, into products with a new use. The result is these throwaway items can be used in paper making as well as plastic products.

Marshall Siao, an architect and director of research and development at Jing Si Pureland, says the paper lunch boxes normally go to the incinerator because it is too difficult to remove the thin plastic coating from the paper underneath. But by removing the plastic coating, the used lunch boxes can be put to a new use in the paper-making process.

"Tzu Chi volunteers strip the thin plastic coating from the paper. The paper, mixed with used newspapers, is then used in making tissue paper."

Cheng Loong, one of Taiwan's leading paper producers, has teamed up with the company. "They are very happy that we get rid of the plastic for them," says Mr. Siao.

Jing Si Pureland also has a corporate partner that can use the plastic to make interlocking pavement tiles which it calls eco-pavers. "This can replace concrete or cement and it is what we call a double loop or zero waste recycling system," says Mr. Siao.

Pictured here is an interlocking paving brick — a recycled "eco product" designed by Jing Si Pureland. Interlocking bricks are a tough paving material made from the plastic membranes that line paper lunch boxes and paper cups. The paper from these products is also put to use. It is sterilized and then recycled to make tissue paper. (Photographer / Peng Wei-Yun)

The company has also eliminated one step in the production process; there is no need for pelletizing the plastic material, thereby saving time and money. The company's floor tiles, called Jing Si eco-pavers, are still under development, but the Hualien Tzu Chi Hospital has made use of them as part of a product test. Mr. Hsiao calculates that every 100 square meters of Jing Si eco-pavers saves four tons of plastic waste from incineration. It could also replace 20 metric tons of concrete and that in turn is equal to reducing nine tons of CO_2. He also calculates that every 100 tons of paper that is recycled can produce 60 tons of tissue paper or 60 tons of packaging material. That amount of paper could save 2,000 10-year-old trees.

Jing Si Pureland also has a line of other products, many of which do not make use of recycled materials but are meant to dovetail with Tzu Chi's efforts to avoid waste and provide relief in disasters. One such product used in relief efforts is a hard plastic foldable bed. Tzu Chi volunteers are encouraged to bring their own bowls and cups as well as chopsticks when they participate in community events or relief efforts. Jing Si has designed collapsible bowls that are lightweight and take up less space. These were part of a presentation to participants at the 25th Conference of the Parties (COP25) to the United Nations Framework Convention on Climate Change in Madrid in December of 2019.

Chapter 12

Getting the Message Out

"One drop of water eventually becomes a river"
— Master Cheng Yen

Da Ai TV

Tzu Chi has an important vehicle to get its message out to the followers and the world at large — a broadcasting platform called Da Ai TV. Set up in 1998, Da Ai TV offers up a steady stream of news, entertainment and messages on the Tzu Chi values, 24 hours a day. It delivers eight hours of fresh programming each day, with the remaining airtime filled with rebroadcasts. All of this is financed by the hard work of the recycling volunteers and generous donations from Tzu Chi followers. One of its key themes is the importance of recycling and the objective of "co-existing with the Earth."

Da Ai TV serves as a window on the Tzu Chi organization and as a communications platform. Many of the recycling volunteers say they became aware of Tzu Chi — and recycling — by way of the television broadcasts which include Master Cheng Yen's lectures on the Tzu Chi ethos such as the need to protect the environment, avoid waste and follow a vegetarian diet. It reinforces the message of the importance of making better use of the Earth's resources. Tzu Chi programs show how volunteers are careful to carry their own cups, bowls and chopsticks with them when they are on the go. They walk or ride a bicycle

Control room technicians monitor a broadcast on Da Ai TV. The television station plays a key role in promoting recycling and protecting the environment. It is part of a virtuous circle that delivers news of Tzu Chi developments, provides global information and promotes the organization's values. It also plays a key role in inspiring people to join Tzu Chi and participate in its activities. (Photographer / Zhan Jin-De)

if possible, and if a car is necessary, the aim is to use it less often. Air conditioning is not exactly frowned upon but limited use is officially encouraged. All of these themes figure prominently in the television station's messages to Tzu Chi's viewers.

"We are a communications platform," says Cheng Mei-ling of Da Ai TV's adult education programming department. The funds earned from recycling are ploughed into the television station, which beams the messages to the volunteers as well as those who might one day join the movement. "It is a virtuous circle," she says.

Shirley Chin, manager of the programming department, describes it this way: "We are the nerve center of Tzu Chi." The objective is to promote healthier lives and Tzu Chi's spiritual values, she adds. "We are very different from other TV stations. We have a different starting point."

As Master Cheng Yen says, the TV station is a broadcaster of virtue and its mission is to purify hearts. "It's not a conventional broadcasting tool."

Da Ai TV's programs include interviews and original reporting on the environment and recycling as well as Buddhism while there also are animated programs for children. Da Ai TV's programs on recycling have introduced the effort to bring ex-prisoners into the recycling effort as volunteers in order to help them find their way back to a productive and meaningful life. During the coronavirus outbreak, the broadcaster had updates on the epidemic, interviews with health practitioners and practical advice on personal hygiene.

"When I started out, I carried a camera around and did shooting and reporting together," says Cheng Mei-ling. She points to her experience reporting on recycling in Pingtung in the early years, making do with limited resources at Tzu Chi. She recalls running alongside the collection trucks as they made their rounds gathering recyclables while she held her camera and shot video. "I never expected this to wield so much influence."

The station leases three satellite transponders — one for its Taiwan viewers, one for those in Malaysia, where Da Ai TV has a large audience, and another for mainland China. At one time Da Ai TV had as many as six transponders though high costs convinced the broadcaster to cut back and accept a more gradual expansion. It has free-to-air broadcasts with a high definition channel and a modest share of Taiwan's fragmented market. In addition to satellite and terrestrial delivery, it has a growing following on Youtube, Facebook, Line (a mobile messaging and social media app widely used in Taiwan, Japan and South Korea), and Wechat (a similar app popular in mainland China) as well as the Da Ai website and the Da Ai TV app.

Officials explain that it was Master Cheng Yen who insisted that Tzu Chi should accept the high cost of running a TV station, which

Jordanian government official Rasheed Aied Rasheed Hijazi, as part of a delegation from the Jordan Hashemite Charity Organization, tours the set of Da Ai TV. The television station reaches a global audience through free-to-air broadcasts, satellite, the Internet and social media apps. (Photo provided by Chen De-Xiong)

has been seen as an essential tool that gradually builds a following. They point to a phrase used by Master Cheng Yen: "One drop of water eventually becomes a river; one grain of sand eventually becomes a tower."

The TV station buys some programs — mostly documentaries — from Japan's NHK and CNN, for example, but it produces about 95% of its programs in-house to cut down on costs. It also benefits from content from many amateur contributors who provide videos, which are edited by the station's staff.

"Many people said TV is so expensive, but the Master insisted we persevere," says Shirley Chin. Recycling covers about 25% of its costs. It also has many volunteers who provide amateur video which Tzu Chi edits. Tzu Chi also gets contributions from some of its 6,000 volunteers in South Africa, for example.

One of its most enduring programs is the Tsaogen Puti — or the Grassroot Enlightened Ones. These are stories about volunteers — ordinary people — who play a critical role in maintaining the recycling operations.

"We are trying to let more people know about Tzu Chi," says James Ho, associate general manager. "And we discovered there are so many good stories drawn from our own volunteers."

One of those who has featured prominently in Da Ai TV's programming is Tsai Kuan, the 101-year-old volunteer who has been a recycling and medical volunteer for more than 30 years. She was a midwife years ago who gradually became interested in Tzu Chi's good works. Tzu Chi has produced a documentary on her extraordinary life as well as a drama.

Da Ai TV has a particularly obvious impact at times of key events such as the Sichuan earthquake in 2008. There was a big response to its programs and calls for donations. Programs geared to other social issues have also had a big impact.

"We did a program on the difficulties of getting water in Gansu," says Ms. Cheng, speaking of the arid province in northwestern China. One memorable program called "The One Centimeter Pencil" focused on life in the relatively poor province, describing the hardships of primary school students who used their pencils down to the stub, or the last centimeter.

"My children have been influenced by stories like these," says Ms. Cheng of the adult education programming operation. "I have two girls — 9 and 12 years old — and they watch these programs. Now they have the right value system."

Even Da Ai TV's weather reports contribute to the Tzu Chi goal of protecting the Earth and raising climate awareness.

"I have an ambition to educate people," says Chi Ming Peng, a weather presenter for Da Ai TV. "Weather presenters are the most

Da Ai TV also provides entertainment programs, most of them its own productions. Shown here is a talk show called "Spiritual Discussions on Selections from the Jing Si Book Store" and recorded before a studio audience. This show's discussion is about the 18th century novel, the Dream of the Red Chamber, a Chinese language literary classic. (Photographer / Xie Xie-Xiang)

trusted and most awarded communicators," he says. He notes that studies by the United Nations Intergovernmental Panel on Climate Change (IPCC), a key force for promoting environmental protection, are technical and seem very far away from the lives of most people. But he adds that he feels he has a responsibility to discuss their conclusions on climate change. "In my weather reports I use one minute to mention the IPCC report."

"Protecting the Earth is my goal," he told a news briefing held on the sidelines of the COP24 meeting in Katowice, Poland, in December 2018. "For example, I teach everybody why I am a vegetarian. It's good for the Earth and food security."

Da Ai TV's staff, combined with the publishing and communications arm of Tzu Chi, has about 600 members. It is organized along the lines of the group's hospitals, which have paid doctors and nurses supplemented by many volunteers. The paid staff do a great deal of volunteer work as well as their regular duties. "They become like volunteers," says Ms. Chin.

Tzu Chi expanded abroad in 2000 with DAAI TV Indonesia. It now has free-to-air programming both in the capital Jakarta and Medan on Sumatra, while cable service allows it to be seen in other markets. With about 200 employees, it broadcasts new programs for three hours a day and then rebroadcasts them over the rest of the programming day from 6 am to midnight. Popular programs are "The Master's Aphorisms," which discusses Master Cheng Yen's more widely quoted sayings, as well as stories about volunteers, talk shows, Tzu Chi-produced documentaries and children's programs. It relies mainly on corporate sponsorships for its funding.

Chapter 13

Global Outreach — Promoting Environmental Protection Through the UN

"Not just words. Demonstrate for all to see"
— Master Cheng Yen

Taiwan has been shut out of many international organizations but that hasn't prevented Tzu Chi from making important contributions to global efforts in support of environmental and other social causes. As a faith-based organization that has been on the frontline of disaster relief and other humanitarian efforts around the world, it has earned a seat at the table when it comes to global health and welfare issues. Global gatherings have given it a platform to present its own experience in environmental protection and sustainable development as well as demonstrate humanistic Buddhism in action. Below are just a few of the areas where Tzu Chi has made its voice heard in national and international humanitarian forums:

- In 2003, Tzu Chi became a member of the National Voluntary Organizations Active in Disaster (NVOAD), joining an association of groups in the United States that have made disaster relief work a priority. NVOAD facilitates partnerships among government agencies, for-profit corporations, foundations, and educational and research institutions to provide more effective delivery of services to disaster-hit communities.

- In 2010, Tzu Chi participated in the annual workshop of the United Nations Commission on the Status of Women, which promotes gender equality and the empowerment of women, sharing the experience of its grassroots movement and demonstrating how it contributes to combating climate change.

- In 2013, Tzu Chi officially joined the United Nations Framework Convention on Climate Change (UNFCCC), an international environmental treaty that went into effect in 1994. The parties to the convention have met annually since 1995 to assess progress in dealing with climate change. Tzu Chi has shared its climate change experience at various sessions of the Conference of Parties (COP), the supreme decision-making body of the framework convention. The key themes of its presentations have been "Co-Exist with the Earth," "DA.AI Technology Compassion in Action" and "Faith-Based Innovation."

- In 2016, Tzu Chi shared its Buddhism in action teachings through the United Nations Population Fund. It was invited to speak at the World Humanitarian Summit, the Buddhist and Catholic Dialogue at the Vatican and the United Nations High Commissioner for Refugees. Tzu Chi has expanded the engagement into policy by joining the International Council of Voluntary Agencies, a global network of organizations that seeks to make humanitarian action more principled and effective.

- In 2018, Tzu Chi engaged in the Sustainable Development Goals Paris Agreement movement, through refugee care, refugee health and safety networking, refugee education, job skill training and disaster response and recovery. Tzu Chi was invited to join the Multi-Faith Advisory Council for agencies of the United Nations.

- In 2019, Tzu Chi officially became a United Nations Environment Programme (UNEP) member and participated in the United

As an international non-governmental organization, Tzu Chi has been invited to play a role at key global gatherings, including those under the United Nations. It attended the 2015 UN Climate Change Conference — where the Paris Agreement was reached — and introduced its recycling concept and practices to a wide audience. Shown here is long-serving U.S. volunteer Debra Boudreaux at a press briefing during that historic meeting in Paris. (Photo provided by Tzu Chi Foundation)

Nations Environment Assembly to share its experience in grassroot movements and the promotion of a plant-based diet.

- In 2020, Tzu Chi registered with the technology transfer mechanism, the Climate Technology Centre and Network.

Tzu Chi has shared its experience with the use of multifunctional foldable beds at the UNFCCC and its track record with pre-cooked ready-to-eat meals at the Commission on the Status of Women. It also demonstrated its eco-chairs and tables at the World Humanitarian Summit as well as DA.AI Technology and Jing Si interlocking bricks at the UN Environment Programme.

"We have made an impact," says Debra Boudreaux, a Tzu Chi commissioner and executive vice president in the U.S., in a reference to programs such as the "bottle-to-blanket" recycling effort. "In 2010, only Tzu Chi was doing these things, but many entrepreneurs have endorsed this concept since then. They see an opportunity in the circular economy."

Ms. Boudreaux, who is also senior representative at the foundation's UN New York/Geneva office, said that India — with its strong textile industry — has been using plastic bottles as a raw material to produce clothing. "Tzu Chi has invited a lot of scholars and recycling volunteers (to observe its recycling program) over the years. They see how this is done and then they implement it. We planted the seeds."

Jan Wolf of Tzu Chi in Frankfurt says there are other areas with potential for replicating the foundation's model. "The plastic waste problem is very serious in Africa and also in many Asian countries which have high rates of population growth and sometimes no recycling systems at all," he says. "As Tzu Chi's recycling system is community-based it could be implemented in any community around the world even if the goods collected cannot be directly used for new products such as clothing."

Mr. Wolf notes that Tzu Chi has had preliminary contacts with Plastic Bank, a social enterprise that works to reduce ocean plastic, on possible cooperation. The Vancouver-based company has set up a successful app-based system to encourage people to collect recyclable goods which are then purchased by Plastic Bank for further processing in more developed countries.

He adds that since joining the Climate Technology Centre and Network (CTCN), Tzu Chi has already teamed up with the UNFCCC's technology transfer arm and other organizations in a series of webinars on climate change, sustainable development and food.

"There is a large focus at the UNFCCC on the circular economy as a way to tackle climate change. We hope that our membership at CTCN will allow Tzu Chi to have an even bigger impact and help combat climate change," he says.

At the COP23 meeting on climate change in Bonn in 2017, a Tzu Chi volunteer explains plans for Ethical Eating Day to promote vegetarianism, expand awareness of climate change and end cruelty to animals. Participants in the Ethical Eating Day campaign pledge to follow a vegetarian diet for one day. COP23 was the 23rd Conference of the Parties, the decision-making body of the UN Framework Convention on Climate Change. (Photo provided by Tzu Chi Foundation)

"Climate change is happening really fast and can only be kept in check if civil society is included, if people understand how important it is to reduce greenhouse gas emissions and have some understanding about their own personal carbon footprint," says Mr. Wolf. "There has to be a policy driven top-down approach which should be complemented by a society-driven bottom-up approach."

Conclusion

For Tzu Chi, "Cherish the Earth" is not just an easy slogan to be tossed about casually; it is part of the ethos of the Buddhist charity which embraces simplicity and eschews waste. It draws on the inspirational message from Master Cheng Yen and a humanistic Buddhism tradition that can motivate tens of thousands to work for the good of humanity and protect the environment. It exhorts followers to remember the 5Rs — refuse, reduce, reuse, repair and recycle. In a sense, it believes in reincarnation for materials as well as the individual soul. Materials that have exhausted their useful life in one form should be permitted to return in another.

Tzu Chi also offers a hopeful message that individuals can make a difference and greatly amplify their impact when they work together. It endorses education to promote a greater understanding of its key guiding principles and their ultimate goals such as the need to promote recycling and limit human damage to the Earth. But it also calls on its followers to take action. It has demonstrated that its "bottom up" approach — a grassroots recycling movement — can work effectively and be applied in multiple settings. Tzu Chi starts with the individual adhering to core principles in his or her own life and slowly influencing others.

Tzu Chi has also found a way to engage with the community and give volunteers a cause they feel they can identify with and in many cases — such as for those who once felt discarded and no longer of value — a new sense of self-worth. Community-based recycling can be seen as an important purpose that all can work towards together.

Tzu Chi volunteers take a selfless approach with the conviction that they are participating in this important community mission. They painstakingly sort materials to maximize value, happy to see the proceeds ploughed back into the cause via the work of Da Ai TV or Tzu Chi's publications. In some cases, Tzu Chi uses funds to enable good works such as charity for the less fortunate. And Tzu Chi pursues its goal of recycling even when there is little direct economic value in its efforts. Its goal is an altruistic one — even without income it contributes to the cause of protecting the environment by reducing the amount of waste that needs to be incinerated or buried. Its actions are taken with the knowledge that the Earth needs a helping hand. Ultimately, Tzu Chi maintains that we all share responsibility for preserving our environment. As Master Cheng Yen says: "We did not inherit the Earth from our parents; we are merely borrowing it from our grandchildren."

Appendix

Year	Tzu Chi Environmental Protection	International Environment
1760		The rise of the Industrial Revolution. The beginning of replacing manual labor with machinery.
1830		Second Industrial Revolution. Humanity enters the age of technology.
1839		"Styrofoam" is invented. These products appear in the market in the 1930s, and become widespread across the globe.
1972		"The United Nations Conference on the Human Environment" makes "The Declaration on the Human Environment." This is the beginning of when national

(Continued)

(Continued)

		governments worldwide formally attach importance to environmental protection work.
1973		First energy crisis.
1978		Second oil crisis.
1985		Vienna Convention for the Protection of the Ozone Layer.
1986		Chernobyl nuclear disaster.
1987		Signing of the "Montreal Protocol." It prohibits the use of CFCs and other industrial aerosols, in order to protect the ozone layer.
1988		United Nations Intergovernmental Panel on Climate Change (IPCC) is formed.
1990	"Use our applauding hands to protect the environment." Tzu Chi volunteers put this into action, turning their own homes into recycling centers, which in turn influence the community.	
1991	Tzu Chi University of Science and Technology carries out various environmental protection works, and implements waste sorting on campus.	

(*Continued*)

	Hualien Tzu Chi Hospital Environmental Protection Society is established, advocating environmental protection activities.	
1992	The second wave of "Creating a Pure Land on Earth," aiming to implement greening works among the population, preserve natural spaces for future generations long-term, advocate environmental and life protection, and cherish all resources on Earth. In response to environmental protection, Hualien Tzu Chi Hospital promotes the campaign of bringing one's own meal utensils at the staff cafeteria, and paper utensils are no longer provided.	The United Nations holds the "Earth Summit." The United Nations Framework Convention on Climate Change (UNFCCC) is passed.

(*Continued*)

(*Continued*)

1994	Tzu Chi implements the use of environmentally friendly utensils across the board.	UNFCCC takes effect.
1996	Typhoon Herb disaster. Promotion of the value of soil and water conservation, to protect mountains and oceans.	
1997	Institutionalization of environmental protection work: social work conferences, learning modes of recycling. Holding visiting programs for environmental protection volunteers, visiting recycling plants, recycled paper factories, incinerators, and so forth, in order to increase their level of understanding of recycling and waste sorting.	The UNFCCC passes the Kyoto Protocol.
1999	921 Earthquake Project Hope lays interlocking bricks, allowing the earth to breathe.	

(Continued)

2000		Third Industrial Revolution. Entering a technological age that relies mainly on nuclear power, computers, smart devices, and so on.
2001	Balancing disaster relief and environmental protection: Tzu Chi disaster assistance uses environmentally friendly eating utensils across the board. Launch of the "Tzu Chi Environmental Education Teachers' Training Program." Training seed-teachers for environmental education, to implement the promotion of environmental education in local communities.	
2003	Establishment of the Tzu Chi International Humanitarian Aid Association. Making efforts in research and development of material goods for disaster relief and the ideal of environmental recycling.	

(Continued)

(*Continued*)

	SARS crisis. Promotion of ethical lifestyle and vegetarianism for the protection of life.	
2005	Promotion of "The Five Transformations on Environmental Protection": in youth, in daily living, in knowledge, in family, and in mind.	The Kyoto Protocol takes effect.
2006	Environmental blankets are released.	International Fashion Trends Research Center announces that "Fast Fashion" will be the trend of development for the garment industry for the next ten years.
2007	"Self-discipline and Courtesy" movement promotes that it is good to have courtesy, and the whole population should reduce their carbon emissions.	
2008	DA.AI Technology Company is founded. Typhoon Morakot disaster. An appeal to let the mountains and forests rest and recuperate.	

(Continued)

	World Food Price Crisis and Financial Crisis. A call to society to conserve food and return to a pure and simple way of life (being industrious and frugal, conserving energy, and reducing carbon emissions). Promotion of charitable agriculture: making efforts in organic farming and revitalizing the original countryside and farmland.	
2010	Twenty Years of Environmental Protection: purity is at the source, refinement of environmental protection.	
2011		The world's population reaches seven billion. Fukushima Daiichi nuclear disaster, brought about by the Tōhoku earthquake in eastern Japan.
2012	DA.AI Technology Company gains Cradle to Cradle® Silver Certification.	

(Continued)

(*Continued*)

	Master advocates the ideal of "Eat 80% full, leave 20% to help others." Tzu Chi promotes the zero food-waste movement.	
2013	First time participating in the United Nations Intergovernmental Panel on Climate Change.	
2015	A press conference is held at the United Nations Paris Climate Conference, to promote Tzu Chi's environmental protection ideals. From her experiences at the Climate Conference, Master laments that the world has "shared understanding and consensus," but does not have "collective action."	United Nations Climate Conference passes the "Paris Agreement," which replaces the Kyoto Protocol.
2016	Launching of the "111 Earth Ethical Eating Day."	

(Continued)

2018		Swedish girl Greta Thunberg launches the "School strike for climate" movement, starting a global upsurge. 24th Conference of the Parties to the United Nations Framework Convention on Climate Change (COP24). CEOs of over 40 fashion labels and brands sign the "Fashion Industry Charter for Climate Action." Mainland China enacts a prohibition of the "Global Waste Trade."
2019	Attainment of NGO observer status for the United Nations Environment Programme. First time participating in the United Nations Environment Assembly (Kenya). Environmental Health Care Cloud Database System goes online to look after the health of volunteers. "Jingsi Fuhui Environmental Protection Interlocking Bricks" are released.	Indonesia, Malaysia, Vietnam, Philippines, Cambodia, Sri Lanka and other countries continue by announcing that they will refuse to accept waste from other countries, and will return illegal waste. Scientists announce that June is the hottest month on record since humans starting recording temperatures over a century earlier. The south of France reaches a maximum temperature of 45.9 C.

(Continued)

(Continued)

2020	Thirty Years of Environmental Protection, Thirty Years of Action, Thirty Years of Sustainability.	

Table created by Tzu Chi Foundation, 2019/10/17

Total Recycling Weight by Tzu Chi in Taiwan for 2019

❙ Statistics for period up until 2019/12/31　❙ Source: Buddhist Compassion Relief Tzu Chi Foundation

Total recycling weight (kg)	Total plastic bottle weight	Total paper weight
81,831,205	3,803,635	39,553,633
Total plastics weight	Total iron weight	Total aluminum weight
6,424,653	8,044,999	903,437
Total scrap metal weight	Total old clothing weight	Total glass weight
561,482	3,851,664	11,151,187
Total copper weight	Total aluminum drink carton weight	Total galvanized steel weight
294,832	2,598,649	290,709
Total plastic bag weight	Total battery weight	Other*
3,689,890	187,199	475,236

*
1. The above statistics are in kg
2. Others include: CD-ROM, cell phones, tablets, computers and fluorescent light bulbs

Chart 01

Conversion Figure of Annual Recycled Paper in Taiwan by Tzu Chi Foundation

❙ Statistics for period up until 2019/12/31 ❙ Source: Buddhist Compassion Relief Tzu Chi Foundation

◆ From 1995 to 2019, Tzu Chi has recycled 1,439,695,935 kg of paper, equivalent to 28,793,918.7 full-grown 20 year-old trees.

50kg of recycled paper = one 20 year-old tree

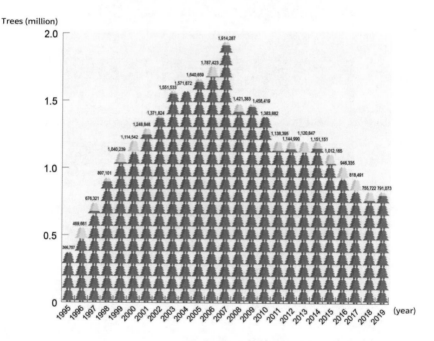

Chart 02

Distribution of Tzu Chi Environmental Protection Volunteers in Taiwan for 2019

❚ Statistics for period up until 2019/12/31 ❚ Source: Buddhist Compassion Relief Tzu Chi Foundation

◆ Total number of Tzu Chi environmental protection volunteers in Taiwan in 2019: 89,585 people

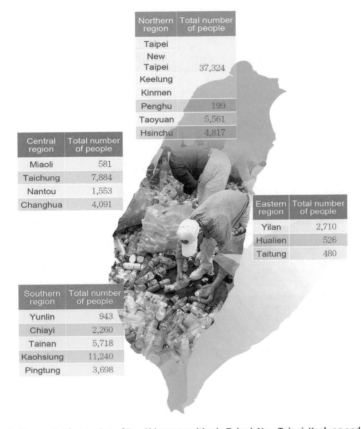

Northern region	Total number of people
Taipei	
New Taipei	37,324
Keelung	
Kinmen	
Penghu	199
Taoyuan	5,561
Hsinchu	4,817

Central region	Total number of people
Miaoli	581
Taichung	7,884
Nantou	1,553
Changhua	4,091

Eastern region	Total number of people
Yilan	2,710
Hualien	526
Taitung	480

Southern region	Total number of people
Yunlin	943
Chiayi	2,260
Tainan	5,718
Kaohsiung	11,240
Pingtung	3,698

*
Due to the mode of operation of Tzu Chi communities in Taipei, New Taipei, Keelung and Kinmen, combined total number of volunteers are provided, as individual statistics are not available.

Chart 03

Distribution of Tzu Chi Environmental Protection Locations around Taiwan for 2019

❚ Statistics for period up until 2019/12/31 ❚ Source: Buddhist Compassion Relief Tzu Chi Foundation

◆ In 2019, Tzu Chi had 273 environmental protection stations and 8,536 community environmental protection points, for a total of 8,809 locations around Taiwan.

Northern region	Environmental protection stations*1	Community environmental protection points*2	Total
Taipei			
New Taipei	35	2,057	2,092
Keelung			(*3)
Kinmen			
Penghu	1	16	17
Taoyuan	24	274	298
Hsinchu	15	188	203

Central region	Environmental protection stations	Community environmental protection points	Total
Miaoli	6	87	93
Taichung	41	699	740
Nantou	7	381	388
Changhua	21	348	369

Eastern region	Environmental protection stations	Community environmental protection points	Total
Yilan	8	123	131
Hualien	4	297	301
Taitung	3	250	253

Southern region	Environmental protection stations	Community environmental protection points	Total
Yunlin	7	250	257
Chiayi	17	598	615
Tainan	27	862	889
Kaohsiung	37	1367	1404
Pingtung	20	739	759

1. Environmental protection station: One of Tzu Chi's community practice centers, it is a site for collecting recycled items by Tzu Chi volunteers at various communities. It is open to the public to understand Tzu Chi and environmental protection, so that they may in turn engage in environmental protection.
2. Community environmental protection point: Under the direction of Tzu Chi volunteers, people in the community who support Tai Chi's ideal of environmental protection collect items for recycling at a specific point and time, at an appropriate location. The items are then sent on to Tzu Chi environmental protection stations to extend their lives.
3. Due to the mode of operation of Tzu Chi communities in Taipei, New Taipei, Keelung and Kinmen, combined total number of volunteers are provided, as individual statistics are not available.

Chart 04

Number of Global Tzu Chi Volunteers and
Number of Environmental Protection Stations/Locations

❙ Statistics for period up until 2019/12/31 ❙ Source: Buddhist Compassion Relief Tzu Chi Foundation

◆ Up until the end of 2019, 532 environmental protection stations and 10,012 community environmental protection points were set up across 19 countries and regions around the globe. A total number of 112,016 environmental protection volunteers dedicated themselves to the mission of environmental, protecting and guarding the Earth through activism.

Asia (12)			
Country/region	Environmental protection stations	Community environmental protection points	Number of environmental protection volunteers
Taiwan	273	8,536	89,585
China	44	323	6,000
Malaysia	154	876	12,349
Indonesia	28	41	600
Philippines	4	16	185
Hong Kong	2	18	180
Thailand	—	3	53
Singapore	1	40	1,100
Brunei	1	1	15
Sri Lanka	0	2	65
Vietnam	—	5	50
Cambodia	—	5	131
Total	509	9,929	110,268

The Americas (4)			
Country/region	Environmental protection stations	Community environmental protection points	Number of environmental protection volunteers
USA	19	9	197
Guatemala	—	1	3
Canada	—	23	710 (participations)
Chile	—	4	20
Total	19	37	930

Oceania (2)			
Country/region	Environmental protection stations	Community environmental protection points	Number of environmental protection volunteers
Australia	2	3	199
New Zealand	1	1	35
Total	3	4	234

Oceania (2)			
Country/region	Environmental protection stations	Community environmental protection points	Number of environmental protection volunteers
South Africa	1	42	584
Total	1	42	584

1 Community environmental protection point: Under the direction of Tzu Chi volunteers, people in the community who support Tai Chi's ideal of environmental protection collect items for recycling at a specific point and time, at an appropriate location. The items are then sent on to Tzu Chi environmental protection stations to extend their lives.
2 Environmental protection station: One of Tzu Chi's community practice centers, it is a site for collecting recycled items by Tzu Chi volunteers at various communities. It is open to the public to understand Tzu Chi and environmental protection, so that they may in turn engage in environmental protection.

Chart 05